Physical Assurance

Navid Asadizanjani • Mir Tanjidur Rahman
Mark Tehranipoor

Physical Assurance

For Electronic Devices and Systems

 Springer

Navid Asadizanjani
University of Florida
Gainsville, FL, USA

Mir Tanjidur Rahman
University of Florida
Gainsville, FL, USA

Mark Tehranipoor
University of Florida
Gainesville, FL, USA

ISBN 978-3-030-62611-2   ISBN 978-3-030-62609-9   (eBook)
https://doi.org/10.1007/978-3-030-62609-9

This Springer imprint is published by the registered company Springer Nature Switzerland AG
The registered company address is: Gewerbestrasse 11, 6330 Cham, Switzerland

Dedicated to:
Our families for their endless support.

Preface

Increased complexity and globalization of the electronics supply chain has made hardware security a critical nationwide necessity. Over the past decade, physical inspection of electronics has grown significantly and is becoming a major focus for chip designers, original equipment manufacturers, and system developers. The long, complex lifespans of modern electronic devices, coupled with their diverse applications, render them increasingly vulnerable to various forms of malicious threats. Efforts of large, global industry and government to address such supply chain security problems involve offering solutions, training, and services. Over the years, the U.S. government has increasingly introduced new programs to analyze and develop relevant solutions. Although much focus is given to the digital domain, analog parameter-based physical assurance, inspection of electronics, and physical fingerprinting are rapidly providing opportunities for unique countermeasures.

This book will detail the principles of invasive, non-invasive, and semi-invasive physical inspection techniques and their roles in hardware assurance from chip to system level. Here, a wide variety of topics will be covered in 7 chapters. Chapter 1 provides an introduction to physical inspection, attack methods, and their application to hardware Trojan and counterfeit chip detection. Chapter 2 presents a case study on the use of physical inspection methods to detect counterfeit electronic components. Chapters 3 and 4 focus on recent advancements in Integrated Circuit (IC) and Printed Circuit Board (PCB) assurance, which is achieved through application of Failure Analysis (FA) tools, image processing, and computer vision techniques. Chapters 5 and 6 describe the security threat to chips imposed by frontside electrical probing attacks and backside optical attacks. Chapter 7 discusses packaging security and integrity assessment with physical inspection. Recent advancements in developing security metrics and countermeasures to protect the device secrets are also presented in these chapters.

Gainesville, FL, USA   Navid Asadizanjani

Gainesville, FL, USA   Mir Tanjidur Rahman

Gainesville, FL, USA   Mark Tehranipoor

June, 2020

Acknowledgments

This work would not have been possible without the help of individuals affiliated with the Florida Institute for Cyber Security (FICS) Research at the University of Florida. We are thankful for their tireless efforts and scientific contributions in different chapters as listed below:

Nathan Jessurun, Chapters 3, 4, and 7,

Leonidas Lavdas, Chapter 2,

Dhwani Mehta, Chapters 2 and 4,

M. Sazadur Rahman, Chapter 1,

John True, Chapter 4,

Nidish Vashistha, Chapter 3,

Huanyu Wang, Chapter 5, and

Chengjie Xi, Chapters 2 and 7.

In addition, we would like to thank Olivia Paradis, Dr. Shayan Taheri, and Jacob Harrison for their thorough review of the book.

Contents

Chapter 1
Physical Inspection and Attacks: An Overview

1.1 Introduction

Embedded and Internet-of-Things (IoT) devices have become an integral part of daily life. Electronic system-on-chips (SoCs) are present in products ranging from consumer smart products (e.g., smartphones and smart appliances), to industrial automation solutions, to military and space applications. The benefits of ubiquitous computing are indisputable, but their proliferation has led to heightened concerns surrounding security and trust. In addition to the software-centric attacks which have been common for decades, modern electronics, especially embedded systems deployed in hostile environments, are vulnerable to *physical attacks*. The same tools and techniques used for advanced failure analysis (FA), defect localization, and reliability analysis of deep sub-micron devices can pose a security threat to hardware if these tools are in the hands of an adversary: probing, fault injection, photon emission, or reverse engineering may all allow attackers to extract secret information or intellectual property (IP) from electronic systems. In the 1990s, physical attacks on smart cards adversely affected the pay-TV industry. During that period, smart cards were widely used for payment applications, and their security was considered state of the art. As counterfeit pay-TV cards siphoned profits away from content providers, security designers had to develop new protection mechanisms against physical attacks.

Tools used for physical attack methods were initially developed to support FA engineers for post-silicon yield analysis and root cause analysis of chips. Over the last two decades, there have been significant improvements in FA tools such as chip polishing, microscopy, probing, focused ion beam (FIB), and X-ray imaging. However, adversaries have also identified how to leverage those same FA methods and tools to attack a chip. Physical attacks have been used for breaching the confidentiality, availability, and integrity of assets on electronic systems (e.g., sensitive information, IP, firmware, and cryptographic keys [34, 48]).

Adverse impacts of physical attacks on electronics range from consumer day-to-day life to national security. For example, sensitive military equipment in enemy hands may result in leakage of information and disclosure of technology details for developing that equipment. During World War II, the Soviet Union manufactured the TU-04 bomber by reverse engineering a captured US B-29 bomber [12]. Researchers have shown that physical attacks enable adversaries to observe a chip's silicon implementation and break into the confidentiality and integrity provided by modern cryptography and security measures. A skilled attacker can use information extracted from a single chip to inject faults, cause denial of service (DoS), or gain remote unauthorized access to a system. For example, an adversary with access to a biometric authentication system can trigger a DoS to convince users to reset their passwords or biometric identities, which can then be spoofed or tampered to access unauthorized information.

In addition to physical attacks, modern SoCs are also vulnerable to attacks by untrusted entities in the supply chain. Over the past two decades, the semiconductor industry business model has shifted from vertical to horizontal. In the horizontal model, original component manufacturers (OCMs) outsource SoC design and fabrication. This allows OCMs to access more advanced offshore design houses and foundries, reduces costs for developing new technologies, and scales down existing IPs. However, this introduces many potentially untrusted entities in the supply chain. Outsourcing design and fabrication renders chips vulnerable to several threats, most notably *hardware Trojans* which entail malicious modifications to the structure and function of a chip [39]. Over the last decade, a number of Trojan detection approaches have been proposed [26, 40, 47, 57]. Existing work suggests FA-based physical inspection methodologies (e.g., reverse engineering and photonic emission analysis) are among the most promising solutions to verify and assess hardware root of trust [3, 43, 53]. Therefore, a good understanding of different physical attack/inspection methods is required to effectively utilize them as trust verification tools and protect a chip's internal assets.

1.2 Physical Inspection and Attacks

Physical attacks are becoming a growing concern in the security community. The primary requirement for a physical attack is unsupervised access to the target hardware. Physical attacks exploit intentionally and unintentionally introduced security vulnerabilities to gain unauthorized access in order to steal, expose, or destroy the hardware's protected assets. In physical inspection, access to the hardware is used to evaluate the trust and assurance of the device. Depending on the nature of sample preparation and invasiveness of the method, physical inspection/attacks can be divided into three classes: (a) non-invasive, (b) invasive, and (c) semi-invasive. In the past, when compared to non-invasive attacks, invasive and semi-invasive attacks were considered as a less concerning threat to security due to higher equipment costs, required expertise, and execution times, along with the fact that the chip would have been destroyed during the inspection or attack. However, in recent years, FA equipment is becoming more advanced, cheaper, and more accessible. Further, focused ion beam (FIB) and scanning electron microscopy (SEM) imaging systems are accessible in many academic/industry labs and can be rented for only a few hundred dollars per hour. Therefore, one should expect to see significant advancements in physical attacks. Equally important, one should also expect advancements in physical inspection-based techniques for effective security and trust verification.

1.2.1   Non-invasive Inspection and Attacks

A non-invasive attack involves extracting assets without leaving footprints or tampering with the packaging or structure of the chip/printed circuit board (PCB) under inspection/attack. There are active and passive approaches for non-invasive attacks. Examples of active non-invasive attacks are fault injection techniques, brute force, and data remanence. In fault injection, the attacker creates an abnormal condition (fault) in the device, to gain unauthorized access to the functionality of the chip. Common fault injection approaches for asset extraction or DoS attacks include voltage glitching and clock glitching [7, 8, 59]. Passive attacks such as side-channel signal analysis have been studied extensively for exposing sensitive data [21, 22, 29]. In recent years, side-channel attacks, e.g., Meltdown and Spectre, have succeeded in extracting cryptographic keys and private data from microprocessors, including the newest Intel and AMD processors [20, 27]. However, such attacks have low success rates in complex ICs, such as multi-core processors. Moreover, several countermeasures against side-channel attacks have been proposed and implemented in modern semiconductor IPs [23, 32, 51]. Also, side-channel signal analysis using transient and quiescent power, delay, and electromagnetic (EM) signals has been widely proposed for trust verification against Trojans [47, 57]. In recent years, non-destructive reverse engineering and optical probing attacks have also been investigated extensively as both defensive and offensive mechanisms.

1.2.2   Invasive Attacks

After invasive attacks are conducted, the chips/devices are destroyed. Typical invasive attacks include reverse engineering, electrical probing, and circuit edit. In an invasive attack, access to the internal components of the hardware is necessary. For example, an IC invasive attack (IUA) requires access to transistors or interconnect layers. A PCB invasive attack requires access to the metal traces or components (e.g., resistors, capacitors, and ICs), which can be exposed by polishing and milling. Due to the destructive nature of invasive inspection/attacks, multiple sacrificial IUAs

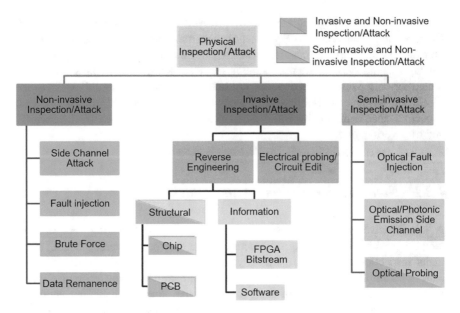

Fig. 1.1 Taxonomy of physical inspection and attacks [36]

may be used. As such, the time and cost of an invasive inspection/attack are greatly influenced by the operator's expertise in sample preparation and the physical attack method.

1.2.2.1 Reverse Engineering

Reverse engineering is the process of analyzing the internal structure (e.g., interconnection and transistors), stored information, and functionality of a chip or PCB. Reverse engineering can be classified as either structural or firmware (see Fig. 1.1). Common reverse engineering tools and instruments include IC soldering/desoldering stations, polishers, plasma etchers, simple chemical labs, high-resolution optical microscopes, X-rays, SEMs, etc. (Fig. 1.2).

A. Chip Reverse Engineering

IC reverse engineering is widely used for understanding the root cause of part failure. It involves five main steps, as shown in Fig. 1.3:

1. **Decapsulation:** Decapsulation, the first step in IC reverse engineering, involves exposing the internal die and the connecting components (e.g., bond wire, ball grid arrays). In non-flip-chips, the internal components are protected with packaging material. In such chips, the package can be removed from either the

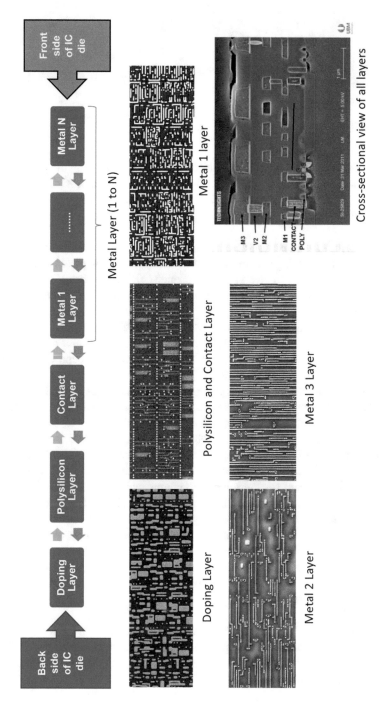

Fig. 1.2 The sequence of layers in an IC along with their cross-sectional view [2, 10, 34]

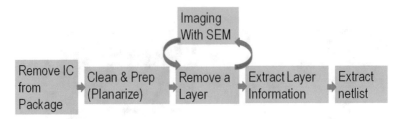

Fig. 1.3 IC reverse engineering process

frontside or backside of the IC. Non-selective means of removing the packaging material include mechanical polishing, computer numerical control (CNC) multitool milling, and wet chemical etching [34, 36, 52]. In flip-chips, the die is covered with a heat sink or lid, which can be removed with a simple knife and hotplate.

2. **Delayering and Deprocessing:** Delayering is the process of removing materials layer by layer for imaging and analysis. Delayering can be completed from either the frontside (interconnect layers) or the backside (silicon substrate). Wet/plasma etching, FIB, or polishing are generally used for layer removal. Iterative physical delayering is one major challenge in IC reverse engineering. Nowadays, an IC consists of several layers of materials, which form interconnects and transistors (see Fig. 1.2). The thickness of each layer varies, which is a major challenge in deprocessing automation. Recently, FIB is used for automated deprocessing due to its advantage of in situ monitoring [33].

3. **Imaging:** After exposing each layer, high-resolution images are collected. In the early days of reverse engineering, optical microscopes were used for image acquisition. Since optical microscopes have a limited field of view, each layer is imaged region by region. Images of each region are then stitched together for a holistic view of the layer. The stitched panorama of each layer is then aligned for netlist extraction. For larger technology nodes, the resolution of an optical microscope is sufficient to determine the features and extract the structure and logic elements of the IC [9, 10, 56]. However, for smaller feature sizes, reverse engineering requires electron microscopy (e.g., a scanning or confocal electron microscope) to acquire high-quality images of chips. In recent years, X-ray synchrotron and ptychography have been used to extract circuit interconnection information from a 14 nm node IC [17]. We note that while the technique was claimed to be non-destructive, the samples for this study were required to be quite small, on the order of tens of microns. Hence, this method is destructive in practice.

4. **Annotation** In this step, all features in the images, such as active region, gates, capacitors, inductors, resistors, vias, contacts, and metal lines, are labeled. Annotation can be manual, i.e., by a subject matter expert (SME), or automated using image processing and computer vision algorithms.

5. **Netlist and Functionality Extraction** Here, different components in the circuit layout are identified and component interconnections are obtained. These components and interconnections are then synthesized into a netlist. Different functional verification and algorithm-based approaches have been proposed for netlist extraction [38]. After the netlist has been extracted, the function of the circuit is analyzed. While netlists and functionality have historically been extracted manually, recent software suites such as ICWorks [2], Pix2Net [1], and Degate [4] automate netlist and functionality extraction.

B. PCB Reverse Engineering

PCB reverse engineering involves identifying all components on the board's front and back surfaces (e.g., resistors, capacitors, ICs) and their interconnections. In two-layered PCBs, components and interconnects are externally visible. However, modern PCBs are trending toward multiple layers, where the majority of the connectivity and structural information is hidden between the layers, i.e., not externally visible. PCB reverse engineering techniques can be destructive or non-destructive. Destructive reverse engineering involves delayering, component removal, and layer-by-layer iterative imaging [34, 41]. It is necessary to collect material thickness, composition, and characteristic information for each layer during destructive delayering. Non-invasive reverse engineering methods include X-ray tomography [6]. During X-ray imaging, image quality and netlist extraction are influenced by material composition, filter, source power, source/detector distance to an object, exposure time, imaging artifact, and tomography algorithm.

C. Bitstream and Firmware Reverse Engineering

A bitstream is a file that contains configuration data for FPGA. SRAM-based FPGAs require external non-volatile memory (NVM). The bitstream is loaded when power is applied. A flash-programmed FPGA uses internal flash memory to hold the bitstream data. Firmware reverse engineering is the process of converting the machine code into a human-readable format. Both bitstream and firmware are stored in non-volatile memory [e.g., read-only memory (ROM), electrically erasable programmable ROM (EEPROM), or flash memory]. Information is stored in the memory cell transistors as electrons. The challenge for reverse engineering memory cells is that any source of energy can potentially disturb the charge distribution and erase the memory content. Prominent NVM extractor tools include scanning probe microscopy, scanning Kelvin probe microscopy, passive voltage contrast (PVC), and scanning capacitance microscopy (SCM) [13]. Probe microscopy uses the direct probing method to extract the charge information. PVC probing involves applying an SEM primary electron beam and detecting the modified secondary beam. Such beam modifications are the result of the presence of an electric field at various locations of the die. Areas with lower charge densities appear brighter in the image.

Then, image processing techniques are used to identify the bit value. SCM involves high-sensitivity capacitance sensors to identify memory cell charges. If bitstream and firmware are encrypted with encryption standards (e.g., DES and AES), the extracted data must be decrypted.

1.3   Electrical Probing and Circuit Edit

IC interconnects carry sensitive information. When the chip is functioning, signals can be read by electrical probing. Such probing is considered a contact-based method for extracting the chip's assets. Circuit editing involves permanently modifying the chip layout connections using a FIB for injecting faults or probing. Electrical probing attacks can be classified into two types: (a) frontside probing [55] through the passivation layer and upper metal layers and (b) backside probing [16] through the silicon substrate.

Wires subjected to probing attacks are called target wires. During probing, the point chosen to serve as the connection between the target wire and the deposited metal contact is called the point of interest (PoI). Desirable PoIs can be identified by reverse engineering. Often, partial reverse engineering is sufficient to extract the data path.

Frontside electrical probing can be challenging due to the large size of the probe tips relative to the size of the available space between wires. To overcome these limitations, attackers typically mill a narrow cavity using a focused ion beam (FIB) to access target wires on lower metal layers. Then, they can build a conducting path without damaging the upper metal layers, as shown in Fig. 1.4. Once the probe-metal layer contact is established, an adversary can extract sensitive information.

Fig. 1.4 (**a**) FIB deposits platinum in the milling cavity to build a conducting path (green) from the target wire. (**b**) The deposited conducting path serves as an electrical pad for the probe contact [54]

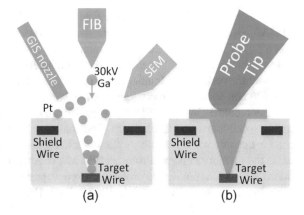

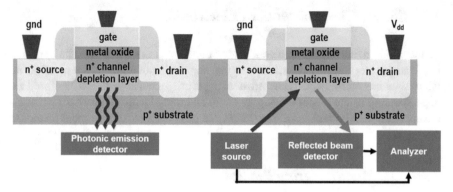

Fig. 1.5 In semi-invasive analysis, photons emitted from transistor switching activity are analyzed. Lasers applied from backside of the chip get modulated due to switching activity of the transistors. The reflected laser is used for laser voltage probing

1.3.1   Semi-invasive Inspection and Attacks

Semi-invasive attacks lie in the gray area between non-invasive and invasive attacks. The main difference between invasive and semi-invasive attacks is that the chip must be powered on in the latter case. Moreover, to launch a semi-invasive inspection/attack, chip decapsulation is sufficient in most cases, as direct contact with the metal layers and transistors is not needed. Therefore, the chip's internal structure remains intact. Semi-invasive attacks are mostly based on optical techniques developed for defect localization. Since the number of interconnect layers increases at the frontside of the chip, optical inspection is performed from the device backside, i.e., a silicon substrate. Optical attacks leverage the transparency of the silicon substrate to near-infrared photons. For asset extraction, photons emitted or modulated due to transistor switching activity are used (see Fig. 1.5). Prevalent forms of semi-invasive inspection/attack include photon emission analysis, laser fault injection, laser voltage probing, laser voltage imaging, and thermal stimulation.

Photon emission analysis and laser voltage techniques (e.g., laser voltage probing, laser voltage imaging [35]) involve monitoring the switching activity of combinational gates and sequential elements. Laser fault injection is applied when setting/resetting a logic gate is required. All optical techniques, excluding photon emission analysis, are active monitoring approaches. Semi-invasive attacks such as photon emission and laser voltage techniques can be non-invasive if the chip package is flip-chip. In the case of flip-chip packaging, backside thinning can be avoided at the cost of lower resolution. Semi-invasive attacks impose a significant threat on the chip's security due to their low cost and short evaluation time.

1.4   Supply Chain of Modern Electronics

In the modern horizontal semiconductor supply chain, several stakeholders are involved in the design and manufacturing steps. Outsourcing different steps of IC design has introduced many trust and security concerns [9]. Therefore, an understanding of the electronic supply chain and manufacturing process facilitates the application of both physical inspection and physical attack/assessment methods.

1.4.1   IC Manufacturing Process

Due to the need for continuous device scaling, designers fit more functionality in a single chip. Integrating the overall functionality of a system of many IPs in a single chip improves speed, power, and area and reduces the development and production cost by minimizing the required silicon area. Such chips are referred to as a system-on-chip (SoC).

The vast majority of mobile and handheld devices contain SoCs, as do many embedded devices. In general, an SoC contains analog components (e.g., radio-frequency receiver, analog-to-digital converter, network interfaces), digital components (e.g., digital signal processing unit, graphics processing unit, central processing unit, serializer-deserializer, cryptographic engine), and memory elements (e.g., RAM, ROM, and flash). Considering the design complexity of modern SoCs and strict project deadlines, it is infeasible for a single design house to complete an entire SoC without outside support [30]. Moreover, the financial cost of building and maintaining a fabrication facility (aka foundry or fab) for modern technology nodes is currently in the multi-billion dollar range. For example, TSMC estimates that their *3 nm* future fab will cost *$20 billion* [50]. Such large initial costs have forced the majority of SoC design houses to turn fabless.

SoC design is an iterative process involving multiple entities, e.g., third-party IP vendors, design service providers, and the design house itself [35] (see Fig. 1.6). The two major phases of SoC design are front end of line (FEOL) and back end of line (BEOL). FEOL processes include design specification, SoC integration, functional verification, design synthesis, and formal equivalency check. BEOL processes include test/debug structure insertion, physical design involving place and route, and design verification.

1.4.1.1   Design Specification

Design specification is the first step of the IC manufacturing process. Here, a design house specifies the high-level requirements and architectural specifications of an SoC. For example, a design house may specify the functionalities it wants to implement in the SoC and a target performance to achieve. To specify the

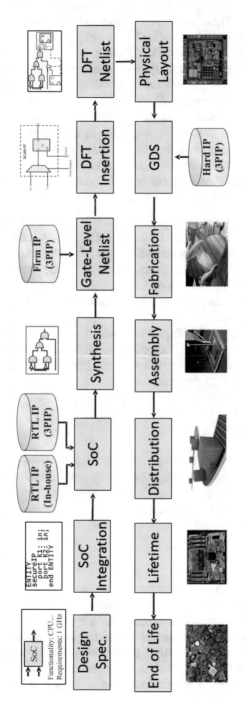

Fig. 1.6 Integrated circuit supply chain [9]

functionalities, the design house identifies a list of hierarchical functional (logic) blocks to implement. These functional blocks may be custom-designed, but a handful of them are pre-designed (either by the design house or by other vendors). Pre-designed functional blocks are called IP cores.

Whether the design house develops an IP in-house or procures it from a third-party IP developer largely depends on factors such as time to market, in-house resources, and cost. For example, a design house specializing in digital design may not want to invest resources in developing an IP involving analog functionality. In another case, a design house with a strict time to market may not have time to develop an IP. In both cases, the design houses may find it more feasible to procure the necessary IPs from third-party vendors (3PIPs).

3PIP acquisition is one of the most critical steps in hardware integration. This acquisition can be accomplished in the following three forms:

- Soft IP cores are delivered as synthesizable register-transfer level (RTL) code written in a hardware description language (HDL), e.g., Verilog or VHDL. Most IPs are procured as soft IPs, as they offer more flexibility than other forms of IP.
- Firm IP cores are delivered as pre-synthesized RTL code represented as a netlist of logic gates and wires. They may use a generic library. Unlike soft IP cores, firm IP does not possess behavioral information. Therefore, firm IPs offer less flexibility compared with soft IPs.
- Hard IP cores are delivered as GDSII representations of a fully placed and routed design. Hard IPs are integrated during the final stages of the design process. They offer the least flexibility, but also a lower cost. For example, most memory and analog mixed-signal IPs are procured as hard IPs.

1.4.1.2 SoC Integration

In this step, IPs developed in-house and procured by third-party vendors are incorporated to generate specifications of the entire SoC. This integration is verified through extensive functional testing. Such functional testing is used to check functional correctness and locate design bugs.

1.4.1.3 Synthesis

Here, abstract RTL specifications are transformed into design implementations in terms of logic gates. An SoC integrator synthesizes RTL descriptions into gate-level netlists based on a target technology library. Synthesis is then performed, at which time the synthesis tool optimizes a design for power, performance, and area. The gate-level netlists are then verified by checking that the synthesized netlists are equivalent to the RTL representations, a process known as formal equivalence checking. At this stage, SoC designers may also integrate firm 3PIP cores into the SoC netlist.

After the synthesis step, the front end of line (FEOL) process ends. Design houses usually keep FEOL in-house, while BEOL can be outsourced or completed in-house as well. The choice of third-party involvement depends on the design house's business model, resources, technological advancement, and return on investment. With an increase in globalization, design houses are trending toward outsourcing BEOL to third-party design service providers [14] more than ever.

1.4.1.4 Design-for-Test

Design-for-test (DFT) refers to the addition of test infrastructures to improve post-silicon SoC testability. Higher testability leads to improved test coverage, quality, and lower costs. DFT enables the IC to be thoroughly tested during fabrication, package assembly, and in-field to ensure its correct functionality. SoC integrators incorporate DFT structures into SoCs to achieve these objectives. However, in many cases, DFT insertion is outsourced to third-party vendors who specialize in designing test and debug structures, e.g., scan, built-in self-test (BIST), and compression/decompression structures [35].

1.4.1.5 Physical Layout

In this phase of the design flow, the gate-level netlist is translated into a physical design layout. Here, each gate is translated into its transistor-level layout. The physical layout also includes transistor placement, wire routing, and placement of clock trees and power grids. At this stage, it is also possible to integrate hard IP cores from vendors into the SoC. After the SoC integrator performs static timing analysis (STA) and power closure, it generates the final layout in GDSII format and sends it to a foundry for fabrication. The generated GDSII file contains layer-by-layer information needed to fabricate the SoC on a silicon wafer.

1.4.1.6 Fabrication

This is the first step in the production phase. As the technology for integrated circuits and SoCs shrinks to deep sub-micron level, the complexity and cost of chip fabrication increase significantly. Therefore, only a few companies can afford to maintain state-of-the-art fabrication facilities. Hence, most design houses have become fabless over the past two decades, meaning they rely on third-party offshore foundries to fabricate their products. SoC designers benefit from state-of-the-art fabrication technologies at low costs, but at the expense of reduced control over product integrity and, therefore, reduced trust in the manufacturing process. The foundry also performs structural/functional tests on the die to detect manufacturing defects, which result from imperfect fabrication processes. The percentage of defect-free chips out of the total manufactured is called yield. Faulty chips are discarded, while the defect-free chips are sent for assembly and packaging.

After fabrication, the foundry sends tested wafers to the assembly line to be cut into die, and dies which pass tests are packaged into complete chips. Advanced assembly operation also includes wafer/die bumping, die placement, solder reflow, underfill, encapsulation, and substrate ball attachment. After these processes are done, assembly performs structural tests to find defects in the chip introduced during the assembly process. After performing these tests, the chips without defects are shipped to distributors or system integrators. Wafer testing and package testing are performed by the foundry and the assembly, respectively. These are mostly structural tests, e.g., automatic test pattern generation (ATPG)-based tests, performed to find defects in the chip introduced during the fabrication and the assembly process. These tests do not necessarily test chip functionality. However, final testing performed during the quality assurance process often focuses on testing chip functionality.

The tested ICs are sent to either distributors or system integrators. Distributors sell ICs in the market. These distributors include original equipment manufacturer (OEM) authorized distributors, independent distributors, Internet-exclusive suppliers, and brokers.

Chips are then combined with other components and subsystems to produce a complete system. This job is typically outsourced to a third-party PCB assembly company, which mounts all the necessary components into one or more PCBs to make the final product. Once the final product is assembled, it is sent to the consumer.

When electronics age or become outdated, they are typically retired and replaced with an upgraded version of the same product or a successor. Proper disposal techniques are advised to extract precious metals and prevent hazardous materials, such as lead, chromium, or mercury, from harming the environment. Proper disposal also ensures that such parts are not maliciously recycled, so they find their way into the supply chain again.

1.4.2 Potential Adversaries

In general, any untrusted entity in the supply chain is a potential adversary. The definition of an attacker is dependent on the threat model considered during the design specification step. A threat model is a procedure to identify the assets and vulnerabilities in any system. A threat model considers the possible dangerous scenarios resulting from vulnerabilities and the attack approaches an adversary could adopt to exploit those conditions. The identity of adversaries will vary from one threat model to another. This section discusses how various supply chain entities could exploit the vulnerability of devices to physical attacks. We will present the design assets and capabilities available to an adversary.

1.4.2.1 Foundry

In hardware security, the offshore foundry is considered a potential antagonist to security and trust issues in the supply chain [31, 35, 47]. The foundry has access to GDSII files used to develop masks for fabricating a chip. Besides, the foundry has access to DFT structures for detecting and analyzing failures in a die. Foundries should be assumed to have access to state-of-the-art failure analysis tools, such as SEM, FIB, laser scanning microscope (LSM), transmission electron microscope (TEM), or advanced sample preparation tools, for defect localization and root cause analysis. An untrusted foundry can also obtain activated chips from the open market or sensitive information from insiders in trusted supply chain entities to aid in reverse engineering chips. Therefore, a rogue foundry may reverse engineer chips to steal IP, overproduce and sell chips on the open market, insert hardware Trojans, or extract secrets.

1.4.2.2 SoC Designers

An SoC designer has access to the soft/hard IP core, knowledge about the functionality of each IP, and functional chips.

It is expected that a rogue designer may have access to DFT structures like the scan chain and synthesis tools. Besides, an SoC integrator should be assumed to know the location and implementation details relevant to security assets as well as the implementation of any security measures in a chip. Therefore, with access to suitable tools and fabricated chips, an SoC integrator can launch physical attacks to reverse engineer IP, insert hardware Trojans, or extract assets.

1.4.2.3 Third-Party IP Vendors and Design Service Providers

In the horizontal supply chain, original equipment manufacturers (OEMs) [30] outsource different IP and design steps from third-party IP vendors and design service providers. Since IP vendor and third-party design service providers have access to the hard/soft IP and SoC design, the threat of hardware Trojan insertion cannot be ruled out.

1.4.2.4 End Users

The threat presented by end users is perhaps the most overlooked concern in hardware security. It is commonly assumed that physical attacks are too expensive and complicated to be carried out by an end user, but advancements in reverse engineering and the availability of FA tools could enable attacks by untrusted users. The research community should revisit this threat.

1.5   Physical Inspection as Assurance Tool

The horizontal supply chain for IC production has drastically reduced manufacturing cost and time to market but has also created many security and trust concerns among different stakeholders in the electronic manufacturing process. Physical inspection is emerging as a possible solution to address trust and security issues.

1.5.1   Detecting Counterfeit Electronics

A counterfeit component is an unauthorized copy or a component produced by an unauthorized contractor. Counterfeit ICs, when used in an electronic system, introduce reliability and security issues. Recycled and remarked components are typical examples of counterfeit hardware. Recycled ICs are chips collected from used electronic systems or PCBs that are repackaged or remarked and then sent back into the supply chain as new chips. The remarking process entails the removal of markings on the package or die and remarking with forged information [15, 49]. An untrusted entity can overproduce chips and sell them on the open market. An unauthorized entity may also sell out-of-spec/defective components.

A malicious party can reverse engineer IP to read the netlist, implementation, and functionality of the IP. Recent automation in different stages of reverse engineering, e.g., automated delayering and imaging [6] and automated netlist extraction [1, 2], have contributed to fast IP extraction. The knowledge about the implementation and functionality of IP can be used to insert backdoors [18] or steal intellectual property [34, 35] to produce clone ICs.

Physical inspection has been widely used for counterfeit detection [15]. Different interior, exterior, and material-based physical inspection methods have been developed to identify different types of counterfeit ICs. These inspection techniques may use an optical microscope, X-ray imaging, EDX analysis, or terahertz imaging to identify remarked, recycled, or out-of-specs ICs.

1.5.2   Localizing and Exposing Device Assets

An adversary can use full-blown or partial reverse engineering to expose assets in a device. Localization of assets with partial reverse engineering mostly relies on semi-invasive physical attack approaches, e.g., photon emission analysis, optical inspection of die, and optical probing [37, 46]. Similarly, an adversary could use a physical attack to access assets through a chip's backside to extract memory contents [28, 42, 45] or bypass cryptographic security [24, 35]. The security threat of physical attack is not limited to key extraction or memory reading. It imposes threats ranging from identity theft to service theft or denial of service, to name only a few.

1.5.3 Detecting Hardware Trojan

Malicious changes in the functionality or parameters of an IC are called hardware Trojans. Such changes could include the addition and deletion of transistors, gates, or interconnects. Parametric changes consist of thinning interconnects, weakening gates, and increasing susceptibility to aging. Hardware Trojans are assumed to be activated under rare conditions, making them difficult to activate and detect them using random input patterns. Test approaches based on functional testing and side-channel fingerprinting have been widely proposed to detecting hardware Trojans at the post-silicon stage. Despite some golden IC-less hardware Trojan detection methods [25, 26, 58], a golden IC is required for most Trojan detection methods. It is assumed that reverse engineering can provide such a golden chip.

Full-blown and partial reverse engineering which have garnered attention in recent years have become accepted by the community as effective for hardware Trojan detection [11, 19, 53]. If it is assumed that a golden layout of a chip is available, machine learning approaches have been demonstrated which compare SEM images of an IC's internal layers/components with the actual chip physical layout. For this technique, decapsulation, delayering, and imaging are required to detect hardware Trojans.

1.6 Conclusions

In the semiconductor industry, physical inspection is becoming a promising solution for hardware trust and assurance verification. In the meantime, physical attacks are a growing concern for device security engineers and OEMs. When a physical attack is combined with algorithm-based attacks, e.g., SAT [44] or remote attack [5], it can impose a significant security threat. In this chapter, we discussed different classes of physical attacks and inspection approaches and presented a comprehensive taxonomy for physical inspection and attacks.

References

1. MicroNet Solutions, Inc., 2015. http://micronetsol.net/pix2net-software/
2. Techinsights.com, 2018. http://www.techinsights.com/
3. Trusted integrated circuits (trust), 2021. https://www.darpa.mil/program/trusted-integrated-circuits
4. Welcome to the Degate Project Website, 2019. http://www.degate.org/
5. Alam, M.M., Tajik, S., Ganji, F., Tehranipoor, M., Forte, D.: Ram-jam: remote temperature and voltage fault attack on FPGAs using memory collisions. In: 2019 Workshop on Fault Diagnosis and Tolerance in Cryptography (FDTC), pp. 48–55. IEEE (2019)
6. Asadizanjani, N., Tehranipoor, M., Forte, D.: Pcb reverse engineering using nondestructive x-ray tomography and advanced image processing. IEEE Trans. Compon. Packag. Manuf. Technol. 7(2), 292–299 (2017)

7. Bar-El, H., Choukri, H., Naccache, D., Tunstall, M., Whelan, C.: The sorcerer's apprentice guide to fault attacks. Proc. IEEE **94**(2), 370–382 (2006)
8. Barenghi, A., Bertoni, G., Parrinello, E., Pelosi, G.: Low voltage fault attacks on the rsa cryptosystem. In: 2009 Workshop on Fault Diagnosis and Tolerance in Cryptography (FDTC), pp. 23–31. IEEE (2009)
9. Bhunia, S., Tehranipoor, M.: Hardware Security: A Hands-on Learning Approach. Morgan Kaufmann (2018)
10. Botero, U.J., Wilson, R., Lu, H., Rahman, M.T., Mallaiyan, M.A., Ganji, F., Asadizanjani, N., Tehranipoor, M.M., Woodard, D.L., Forte, D.: Hardware trust and assurance through reverse engineering: a survey and outlook from image analysis and machine learning perspectives. arXiv preprint: 2002.04210 (2020)
11. Courbon, F., Loubet-Moundi, P., Fournier, J.J., Tria, A.: A high efficiency hardware trojan detection technique based on fast sem imaging. In: 2015 Design, Automation & Test in Europe Conference & Exhibition (DATE), pp. 788–793. IEEE (2015)
12. Curtis, S.K., Harston, S.P., Mattson, C.A.: The fundamentals of barriers to reverse engineering and their implementation into mechanical components. Res. Eng. Des. **22**(4), 245–261 (2011)
13. De Nardi, C., Desplats, R., Perdu, P., Beaudoin, F., Gauffier, J.L.: Oxide charge measurements in eeprom devices. Microelectron. Reliab. **45**(9–11), 1514–1519 (2005)
14. Fuller, D.B.: Chip design in china and india: multinationals, industry structure and development outcomes in the integrated circuit industry. Technol. Forecast. Soc. Chang. **81**, 1–10 (2014)
15. Guin, U., DiMase, D., Tehranipoor, M.: Counterfeit integrated circuits: detection, avoidance, and the challenges ahead. J. Electron. Test. **30**(1), 9–23 (2014)
16. Helfmeier, C., Nedospasov, D., Tarnovsky, C., Krissler, J.S., Boit, C., Seifert, J.P.: Breaking and entering through the silicon. In: Proceedings of the 2013 ACM SIGSAC Conference on Computer & Communications Security, pp. 733–744 (2013)
17. Holler, M., Guizar-Sicairos, M., Tsai, E.H., Dinapoli, R., Müller, E., Bunk, O., Raabe, J., Aeppli, G.: High-resolution non-destructive three-dimensional imaging of integrated circuits. Nature **543**(7645), 402 (2017)
18. Hoque, T., Wang, X., Basak, A., Karam, R., Bhunia, S.: Hardware trojan attacks in embedded memory. In: 2018 IEEE 36th VLSI Test Symposium (VTS), pp. 1–6. IEEE (2018)
19. Kimura, A., Scholl, J., Schaffranek, J., Sutter, M., Elliott, A., Strizich, M., Via, G.D.: A decomposition workflow for integrated circuit verification and validation. J. Hardw. Syst. Secur. **4**, 1–10 (2020)
20. Kocher, P., Horn, J., Fogh, A., Genkin, D., Gruss, D., Haas, W., Hamburg, M., Lipp, M., Mangard, S., Prescher, T., et al.: Spectre attacks: exploiting speculative execution. In: 2019 IEEE Symposium on Security and Privacy (SP), pp. 1–19. IEEE (2019)
21. Kocher, P., Jaffe, J., Jun, B.: Differential power analysis. In: Annual International Cryptology Conference, pp. 388–397. Springer (1999)
22. Kocher, P.C.: Timing attacks on implementations of diffie-hellman, RSA, DSS, and other systems. In: Annual International Cryptology Conference, pp. 104–113. Springer (1996)
23. Kocher, P.C., Jaffe, J.M., Jun, B.C.: Prevention of side channel attacks against block cipher implementations and other cryptographic systems (2010). US Patent 7,787,620
24. Krämer, J., Nedospasov, D., Schlösser, A., Seifert, J.P.: Differential photonic emission analysis. In: International Workshop on Constructive Side-Channel Analysis and Secure Design, pp. 1–16. Springer (2013)
25. Lamech, C., Rad, R.M., Tehranipoor, M., Plusquellic, J.: An experimental analysis of power and delay signal-to-noise requirements for detecting trojans and methods for achieving the required detection sensitivities. IEEE Trans. Inf. Forensics Secur. **6**(3), 1170–1179 (2011)
26. Li, M., Davoodi, A., Tehranipoor, M.: A sensor-assisted self-authentication framework for hardware trojan detection. In: 2012 Design, Automation & Test in Europe Conference & Exhibition (DATE), pp. 1331–1336. IEEE (2012)
27. Lipp, M., Schwarz, M., Gruss, D., Prescher, T., Haas, W., Fogh, A., Horn, J., Mangard, S., Kocher, P., Genkin, D., et al.: Meltdown: reading kernel memory from user space. In: 27th {USENIX} Security Symposium ({USENIX} Security 18), pp. 973–990 (2018)

28. Lohrke, H., Tajik, S., Krachenfels, T., Boit, C., Seifert, J.P.: Key extraction using thermal laser stimulation. IACR Trans. Cryptogr. Hardw. Embed. Syst. **2018**, 573–595 (2018)
29. Longo, J., De Mulder, E., Page, D., Tunstall, M.: SoC it to EM: electromagnetic side-channel attacks on a complex system-on-chip. In: International Workshop on Cryptographic Hardware and Embedded Systems, pp. 620–640. Springer (2015)
30. Macworld: Where are apple products made? https://www.macworld.co.uk/feature/apple/where-are-apple-products-made-3633832/
31. Nahiyan, A., Xiao, K., Yang, K., Jin, Y., Forte, D., Tehranipoor, M.: Avfsm: a framework for identifying and mitigating vulnerabilities in FSMs. In: Proceedings of the 53rd Annual Design Automation Conference, pp. 1–6 (2016)
32. Popp, T., Mangard, S.: Masked dual-rail pre-charge logic: Dpa-resistance without routing constraints. In: International Workshop on Cryptographic Hardware and Embedded Systems, pp. 172–186. Springer (2005)
33. Principe, E.L., Asadizanjani, N., Forte, D., Tehranipoor, M., Chivas, R., DiBattista, M., Silverman, S., Marsh, M., Piche, N., Mastovich, J.: Steps toward automated deprocessing of integrated circuits. In: ISTFA 2017: Proceedings from the 43rd International Symposium for Testing and Failure Analysis, p. 285. ASM International (2017)
34. Quadir, S.E., Chen, J., Forte, D., Asadizanjani, N., Shahbazmohamadi, S., Wang, L., Chandy, J., Tehranipoor, M.: A survey on chip to system reverse engineering. ACM J. Emerg. Technol. Comput. Syst. **13**(1), 1–34 (2016)
35. Rahman, M.T., Rahman, M.S., Wang, H., Tajik, S., Khalil, W., Farahmandi, F., Forte, D., Asadizanjani, N., Tehranipoor, M.: Defense-in-depth: a recipe for logic locking to prevail. Integration (2020)
36. Rahman, M.T., Shi, Q., Tajik, S., Shen, H., Woodard, D.L., Tehranipoor, M., Asadizanjani, N.: Physical inspection & attacks: new frontier in hardware security. In: 2018 IEEE 3rd International Verification and Security Workshop (IVSW), pp. 93–102. IEEE (2018)
37. Rahman, M.T., Tajik, S., Rahman, M.S., Tehranipoor, M., Asadizanjani, N.: The key is left under the mat: on the inappropriate security assumption of logic locking schemes. In: Conference on IEEE International Symposium on Hardware Oriented Security and Trust (HOST) (2020)
38. Rematska, G., Bourbakis, N.G.: A survey on reverse engineering of technical diagrams. In: 2016 7th International Conference on Information, Intelligence, Systems & Applications (IISA), pp. 1–8. IEEE (2016)
39. Salmani, H., Tehranipoor, M., Karri, R.: On design vulnerability analysis and trust benchmarks development. In: 2013 IEEE 31st International Conference on Computer Design (ICCD), pp. 471–474. IEEE (2013)
40. Salmani, H., Tehranipoor, M., Plusquellic, J.: A layout-aware approach for improving localized switching to detect hardware trojans in integrated circuits. In: 2010 IEEE International Workshop on Information Forensics and Security, pp. 1–6. IEEE (2010)
41. Shen, H., Rahman, M.T., Asadizanjani, N., Tehranipoor, M., Bhunia, S.: Coating-based PCB protection against tampering, snooping, em attack, and x-ray reverse engineering. In: ISTFA 2018: Proceedings from the 44th International Symposium for Testing and Failure Analysis, p. 290. ASM International (2018)
42. Stellari, F., Song, P., Villalobos, M., Sylvestri, J.: Revealing SRAM memory content using spontaneous photon emission. In: 2016 IEEE 34th VLSI Test Symposium (VTS), pp. 1–6. IEEE (2016)
43. Stellari, F., Song, P., Weger, A.J., Culp, J., Herbert, A., Pfeiffer, D.: Verification of untrusted chips using trusted layout and emission measurements. In: 2014 IEEE International Symposium on Hardware-Oriented Security and Trust (HOST), pp. 19–24. IEEE (2014)
44. Subramanyan, P., Ray, S., Malik, S.: Evaluating the security of logic encryption algorithms. In: 2015 IEEE International Symposium on Hardware Oriented Security and Trust (HOST), pp. 137–143. IEEE (2015)
45. Tajik, S., Lohrke, H., Seifert, J.P., Boit, C.: On the power of optical contactless probing: Attacking bitstream encryption of FPGAs. In: Proceedings of the 2017 ACM SIGSAC Conference on Computer and Communications Security, pp. 1661–1674. ACM (2017)

46. Tajik, S., Nedospasov, D., Helfmeier, C., Seifert, J.P., Boit, C.: Emission analysis of hardware implementations. In: 2014 17th Euromicro Conference on Digital System Design, pp. 528–534. IEEE (2014)
47. Tehranipoor, M., Koushanfar, F.: A survey of hardware trojan taxonomy and detection. IEEE Des. Test Comput. **27**(1), 10–25 (2010)
48. Tehranipoor, M., Wang, C.: Introduction to Hardware Security and Trust. Springer Science & Business Media (2011)
49. Tehranipoor, M.M., Guin, U., Forte, D.: Counterfeit integrated circuits. In: Counterfeit Integrated Circuits, pp. 15–36. Springer (2015)
50. The Inquirer: TSMC says 3 nm plant could cost it more than $20bn. https://www.theinquirer. net/inquirer/news/3018890/tsmc-says-3nm-plant-could-cost-it-more-than-usd20bn (2017). From the original on 12 Oct 2017
51. Tiri, K., Verbauwhede, I.: A logic level design methodology for a secure DPA resistant ASIC or FPGA implementation. In: Proceedings Design, Automation and Test in Europe Conference and Exhibition, vol. 1, pp. 246–251. IEEE (2004)
52. Torrance, R., James, D.: The state-of-the-art in ic reverse engineering. In: International Workshop on Cryptographic Hardware and Embedded Systems, pp. 363–381. Springer (2009)
53. Vashistha, N., Lu, H., Shi, Q., Rahman, M.T., Shen, H., Woodard, D.L., Asadizanjani, N., Tehranipoor, M.: Trojan scanner: detecting hardware trojans with rapid SEM imaging combined with image processing and machine learning. In: ISTFA 2018: Proceedings from the 44th International Symposium for Testing and Failure Analysis, p. 256. ASM International (2018)
54. Wang, H., Shi, Q., Forte, D., Tehranipoor, M.M.: Probing assessment framework and evaluation of antiprobing solutions. IEEE Trans. Very Large Scale Integr. VLSI Syst. 1–14 (2019). https:// doi.org/10.1109/TVLSI.2019.2901449
55. Wang, H., Shi, Q., Nahiyan, A., Forte, D., Tehranipoor, M.M.: A physical design flow against front-side probing attacks by internal shielding. IEEE Trans. Comput. Aided Des. Integr. Circuits Syst. **39**, 2152–2165 (2019)
56. Wilson, R., Asadizanjani, N., Forte, D., Woodard, D.L.: Histogram-based auto segmentation: a novel approach to segmenting integrated circuit structures from SEM images. arXiv preprint: 2004.13874 (2020)
57. Xiao, K., Forte, D., Jin, Y., Karri, R., Bhunia, S., Tehranipoor, M.: Hardware trojans: lessons learned after one decade of research. ACM Trans. Des. Autom. Electron. Syst. **22**(1), 1–23 (2016)
58. Xiao, K., Forte, D., Tehranipoor, M.M.: Efficient and secure split manufacturing via obfuscated built-in self-authentication. In: 2015 IEEE International Symposium on Hardware Oriented Security and Trust (HOST), pp. 14–19. IEEE (2015)
59. Xiao, Y., Zhang, X., Zhang, Y., Teodorescu, R.: One bit flips, one cloud flops: cross-vm row hammer attacks and privilege escalation. In: 25th {USENIX} Security Symposium ({USENIX} Security 16), pp. 19–35 (2016)

Chapter 2
Counterfeit Detection and Avoidance with Physical Inspection

2.1 Introduction

The supply chain of integrated circuits (ICs) has changed significantly over the last two decades. Due to the high demand for smaller technologies, reduced manufacturing cost, and shortened time to market, more ICs are fabricated in offshore foundries. Though IC manufacturing's globalization has reduced the fabrication costs and accelerated SoC development, the original component manufacturers (OCMs) have lost control over the supply chain. Therefore, vulnerabilities such as counterfeit ICs and hardware Trojans are introduced in the supply chain [24, 58, 60, 61, 61, 70] and become a significant source of reliability and security issues in the present-day electronic supply chain. Many industrial sectors, including automotive, healthcare, communication, Internet-of-Things (IoT), and even military systems, are adversely affected by this significant security concern. The hazard presented by the untrusted supply chain is especially concerning in the context of safety-critical systems, e.g., automated car, aircraft, smart grids, medical equipment, and telecommunication.

It is not unlikely that counterfeit ICs contain hardware Trojans, malicious modifications in circuit design. Any possibility of hardware Trojan insertion violates the root of trust in the device [46]. Other impacts of counterfeit ICs include sales, reputation, and replacement costs to intellectual property (IP) holders. It is also one of the major causes of lost tax revenue and reduced incentive to develop new products and ideas, impacting job creation and employment. In one recent reported counterfeit incident, 30 people were convicted of illegally distributing counterfeit Cisco equipment with the intent of selling to the US Department of Defense [19, 38, 64]. The devices were planned to be installed in Iraq in Marine Corps networks used for security systems and for transmitting troop movements and relaying intelligence to command centers. The severity of the counterfeit issue is clearly understood if the percentage of counterfeit IC used in different market segments is analyzed [2] (see Table 2.1). According to IHS [2], reported cases of

Table 2.1 Top 5 counterfeit ICs used in different market segments

Top part type reported in counterfeit incidents	Industrial market	Automotive market	Consumer market	Wireless market	Wired market	Compute market	Other
Analog IC	14%	17%	21%	29%	6%	14%	0%
Microprocessor IC	4%	1%	4%	2%	3%	85%	0%
Memory IC	3%	2%	13%	26%	2%	53%	1%
Programmable logic IC	30%	3%	14%	18%	25%	11%	0%
Transistor	22%	12%	25%	8%	10%	22%	0%

counterfeit component (both ICs and PCBs) increased fourfold in the time period of 2009–2012 [24], which depicts the severity of counterfeit issues.

The main factor boosting the growth of the counterfeit industry is globalization. Counterfeiters have taken advantage of the complex global supply chain and found ways to replace original components with low-quality counterfeits for profit. As a result, entities like government, end users, and system integrators are unknowingly buying low-quality products with a lower expected lifetime at the same cost.

Although counterfeit IC industry forms a relatively small part of the overall counterfeit business, the ramifications of a counterfeit IC device failure have a colossal impact both in the global supply chain and also to the national security system, especially for those countries which are entirely dependent on import of ICs to meet their domestic demands. Hence, counterfeits are a critical reliability and security concern that must be addressed. According to the Senate Armed Services, the number of counterfeit components found in the US Air Force had surpassed 1 million, underscoring the seriousness of the problem [15].

Physical inspection can be useful for verifying the authenticity of the suspected counterfeit devices [51]. The process of counterfeit chip detection with physical inspection is considered a time-consuming approach. Besides, involvement of highly skilled subject matter experts (SME) makes the entire process costly. The automation in counterfeit IC detection, through the exploration of computationally scalable image processing-based techniques and computer vision, can facilitate in authenticity verification of large volume of ICs while only needing reasonably low-cost equipment.

2.2 Counterfeit Taxonomy

Counterfeit components are defined as those that:

- Do not conform to OCM designs or performance standards
- Are unauthorized copies

- Are produced by unauthorized contractors
- Are defective, out-of-specification, or recycled products that are sold as new
- Have incorrect markings or forged documentation

All these scenarios need to be considered by a counterfeit IC taxonomy. Figure 2.1 presents the taxonomy of counterfeit ICs [24].

2.2.1   IP Overuse

The IP owner and OCM are interested in providing a valuable product to the consumer by continuously improving the IP. Therefore, the IP owner aims to prevent IP loss to the competitors and IP users. The PCB integrator or OCM acquires the right from IP owner to integrate third-party IP (3PIP) in a pre-defined number of chips. A rogue PCB manufacturer may produce a larger quantity of chips and reveal a lesser quantity to the IP owner, which reduces the licensing cost for the PCB integrator. Such deception is possible because IP owners have no definite way to check/verify the number of chips that were fabricated with their IPs. This causes loss of business for the IP owners [4].

2.2.2   IP Piracy

IP piracy is possible when the PCB designer illegally produces copies/clones of the legal purchased 3PIP core and sells them to other PCB designers. Another scenario is when the PCB designer slightly changes the RTL of an existing IP and claims the stolen IP as their own. However, the PCB manufacturers themselves could be victims of IP piracy. A dishonest design-for-test (DFT) vendor could sell parts of the PCB designer's firm IP to other PCB manufacturers [24, 59]. Untrusted foundries could also sell illegal Gerber files. Similar to PCB, an untrusted foundry has access to the GDSII file of the IC. Therefore, an offshore foundry can steal the IP. A competitor could reverse engineer the chip and remove watermarks implemented in the IP to claim the IP as their own. IP piracy may also occur for an IP implemented in FPGA or cloud server.

2.2.3   Overproduction

When chips are produced in excess of a foundry's contract, they are termed "overproduced." Overproduction is also possible when the dishonest foundries report a false low yield. These illegally produced chips would then be sold under a different IP owner's name. Overproduction causes financial loss to design houses

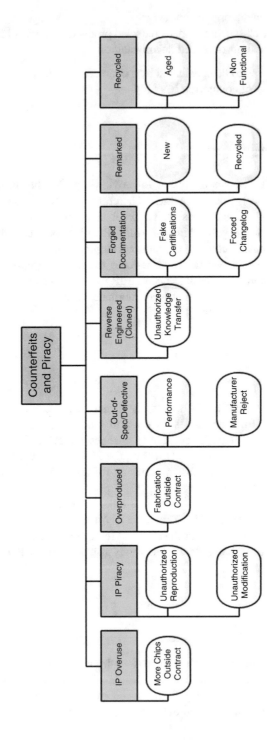

Fig. 2.1 Counterfeit and piracy taxonomy

that have invested in the research and development of their products. Apart from financial loss, a significant concern is the reliability of these overproduced ICs since they may or may not have been tested. If these find places in critical infrastructure, they could inflict damage and put people at risk of harm. Overproduction could also hurt the reputation of the OCM.

2.2.4   Out-of-Spec Parts

Tests are performed after components are manufactured to ensure reliability and functionality. Many manufactured ICs will be defective and will fail these tests. Ideally, defective components are either destroyed or downgraded to a different performance class. However, these parts can be marked as passing and sold in an open market by unscrupulous supply chain entities. This increases the risk of failure where these ICs are used.

2.2.5   Reverse Engineering

Reverse engineering is a process that could be performed for honest or dishonest intentions. For dishonest intentions, the primary motivation is to create a copy of the existing PCB. A company's competitors often perform this type of reverse engineering. X-ray or terahertz imaging could be used to gather information about the subsurface layers of a PCB. Next automated visual inspection, image processing, and machine learning could create images of a PCB, segment them, and generate Gerber files.

2.2.6   Forged Documentation

Each PCB contains a document that summarizes functionality and describes components used on the PCB. By modifying this documentation, these systems could be misrepresented and sold even if they may be defective. It is challenging to verify the authenticity of such documents.

2.2.7   Remarked

PCBs and ICs contain markings that uniquely identify them and their functionality. These markings could contain various information such as the date code, the lot code, manufacturer's identification, and the manufacturing country. These markings identify the origin of a component and give users information on how to handle them [17]. Based on these markings, the component's cost could also vary. A dishonest manufacturer could remark components to sell them at a higher price.

2.2.8 Recycled

A recycled PCB or IC is reclaimed or recovered from a system and then visually restored to appear as new [26, 27, 71]. These electronic components may have reliability or performance issues because they have already aged. Further, the processes by which they are recovered may include high temperatures or aggressive physical removal which could further damage parts.

Recycling causes financial losses to multiple parties. The reputations of companies could suffer if consumers have bad experiences. Recycled electronics can cause severe damages if they fail when they are used in critical infrastructure. It has become of utmost importance to identify these counterfeit systems.

2.3 Counterfeit Detection

Counterfeit detection techniques are an area of active research, but most currently available techniques involve a subject matter expert (SME) which makes the process error-prone, time-consuming, costly, and un-scalable to industrial development volumes. Some of the examples of counterfeit are shown in Figs. 2.2 and 2.3. Counterfeit detection can be broadly classified into the following categories:

1. **Physical inspection:** Physical inspection is a promising approach which relies heavily on computer vision, image processing, and machine learning. Physical inspection can be either destructive or non-destructive. Emerging and established physical inspection approaches are explained in detail in the sections below.

Fig. 2.2 Example of cloned PCB [63]

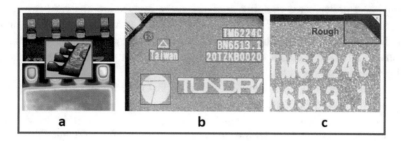

Fig. 2.3 Possible counterfeit candidates: (**a**) bent LEDs, (**b**) marking and localization displacement, and (**c**) texture analysis

2. **Age-based fingerprinting:** This is used to fingerprint a device using some characteristic that varies over a circuit's lifetime. One example of such a technique is creating a signature using electromagnetic (EM) or radio-frequency (RF) emissions. The features are extracted, recorded, and matched against a database if required. However, this technique is not widely used because of its expense and the high precision required for these measurements. DNA markings on each component are mandated by the US Department of Defense (DoD) to trace components through the supply chain [14, 25, 39].

2.4   Physical Inspection-Based Counterfeit Detection

This section concentrates on counterfeit detection using physical inspection techniques. Physical inspection is the most widely accepted approach due to its high accuracy. These techniques use image processing and computer vision and are motivated from quality control and manufacturing automation used in industrial process [10, 22, 43]. In the domain of IC and PCB reverse engineering, many of these approaches have found their applications [6, 47, 65]. This class of techniques leverages defects present in counterfeit ICs. Physical inspection can be performed in a non-destructive manner, in automated or semi-automated fashion based on image processing, computer vision, and machine learning algorithms. These techniques help save time, are cost-effective, and do incur the human error associated with SMEs.

2.4.1   Visual Inspection

To evaluate and authenticate the PCBs and ICs at the surface, volumetric, or material levels, visual inspection is used [8]. Visual inspection depends on different imaging modalities such as X-ray, visible light, or SEMs. Once the images are collected through the pre-selected imaging modalities, the images are analyzed for defects or tampering. In the past, SMEs have done this, but an inspection by humans is tedious and prone to errors. These disadvantages lead to the development of automated visual inspection relying heavily on computer vision, image processing, and machine learning instead of SMEs.

One easy technique is to use cameras or optical microscopes for imaging [21]. This could be used to detect irregular component placement, which may indicate that a component has been tampered, or for detecting counterfeits or damaged components. Damage may include bent or corroded pins, scratches, absence of a logo, marking localization, or irregular texture. In the past, golden ICs were used to match devices under test (DUT) to find probable counterfeits [21]. In such techniques, image processing algorithms such as local binary patterns and active contour detection are used. For further in-depth analysis, multiple two-dimensional images are combined to construct a three-dimensional model for further analysis.

2.4.2   X-Ray Imaging

X-rays have been used for non-destructive reverse engineering of PCBs and chips [5]. X-rays could be 2D (X-ray radiography) or 3D (X-ray tomography). X-rays are commonly used to evaluate internal defects that could exist because of counterfeiting or reverse engineering. In previous work, the authors have used X-ray imaging for defect detection with the main aim of counterfeit detection [16]. Image processing algorithms, such as local binary patterns, were used to extract features from X-ray images. Later, machine learning-based classifiers, such as support vector machine (SVM) and deep learning algorithms, were constructed. While this technique sounds promising, it has several disadvantages. These disadvantages include:

- The reconstructed image could have artifacts,
- It requires several pre- and post-processing stages for reconstruction,
- X-ray imaging tools require expensive maintenance,
- X-ray imaging is slow compared to other imaging modalities

2.4.3   Terahertz Imaging

Terahertz imaging is a rising technique for visualizing internal layers of PCBs. The range of terahertz waves starts at the end of the infrared spectrum and extends to the start of the microwave spectrum. When terahertz waves pass through PCBs, some are partially reflected, while others are partially absorbed. Based on the terahertz spectrum's unique spectral fingerprints, it is possible to identify materials used in components and PCBs. Along with material inspection, terahertz can also be

Table 2.2 Comparison between different imaging modalities

X-ray imaging	Terahertz imaging
Extremely expensive	Cost-effective
Time-consuming	Fast
Ionizing	Non-ionizing
(Photon energy 100 eV to 100 KeV)	(Photon energy 0.4 eV to 40 meV)
Difficult to detect black-coated components	Can detect black-coated components

Fig. 2.4 Steps for image analysis

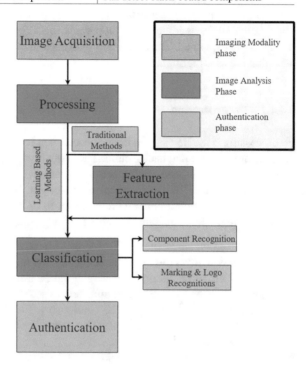

used for electronic components' characterization [3]. A quick comparison between terahertz imaging and X-ray imaging is shown in Table 2.2.

2.4.4   Image Analysis

PCB/IC images are acquired based on the chosen imaging modality. These images then go through a process of algorithms such as image processing algorithms, computer vision algorithms, and machine learning algorithms. The images are analyzed using these algorithms to detect any defects, marking localize logos, and flag malicious components. Figure 2.4 presents a flowchart of the Automatic Bill of Materials (AutoBoM) process for PCB assurance. This process is further divided into three phases: image acquisition (for full details, see section above), image analysis, and authentication. The image analysis phase is the most intricate of the three phases and is explained below:

Preprocessing, Feature Extraction, and Classification Image processing depends on the imaging modality, the quality and resolution of the acquired images, and also data availability. There is limited research on counterfeit PCB detection using the image processing approach, even though image processing is widely used in object detection and classification tasks [40]. Therefore, a comprehensive explanation for preprocessing, feature extraction, and classification is given below. For a better understanding of the approach, algorithms related to these steps are included and explained. The challenges with each of these algorithms for counterfeit detection are also highlighted.

2.4.4.1 Preprocessing

Preprocessing is the first step in image processing. It removes unnecessary information, including noise and background. The noise usually comes from the instrument used for image acquisition. Artifacts in X-ray are an example [29], as are reflections on the surface of PCBs from optical microscopes. Background is defined as the image area that does not contain components to be inspected, traces, or vias. The following are methods that are available for image preprocessing.

1. **Filter-Based Noise Removal:** Traditional filters, including Gaussian filter [31], differentiation filters [50], median filter [30], and mean filter [32], are capable and quite efficient in capturing local neighborhood information [44]. They preserve component features so that essential features can be extracted to improve the performance in further steps. The challenge is the filter size selection, usually defined from the empirical analysis that makes the inspection process difficult to be fully automated.
2. **Background Subtraction:** Background subtraction is a simple but efficient preprocessing method that is useful for removing unimportant information from a PCB image [44]. For example, inspection for extra components can be achieved by counting the remaining areas after background subtraction. Existing background subtraction methods used in hardware inspection include frame/image averaging [1], frame/image differencing [20, 69], mean filtering [32], eigen background subtraction [44], etc. One major drawback of background subtraction is that many methods are influenced by lighting conditions.
3. **Morphological Operations:** Morphological operations operate on the shapes of regions in images. Popular techniques include region dilation, region erosion, skeletonization, and others [18, 37]. Morphological operators are used for refining segmented areas and improve the performance of feature extraction and localization steps. Similar to filter size selection, the parameters of operators may vary between applications.
4. **Thresholding in 3D Depth Map:** The simplest method for locating components is to measure three-dimensional information using a sensor. It is just a height-

thresholding process that slices out components that are not in the same plane as the board. In this way, anything except traces, vias, and text can be segmented out. In the case that there are multiple components in the same height level, more complex algorithms must be needed. The accuracy of this method depends on the resolution of the measurement instrument used. The greater the z resolution, the better thresholding results will be. However, for many components, even after thresholding, segmentation has to be performed.

Challenges in Preprocessing A significant drawback is that preprocessing is heavily dependent on the acquired image. Also, the algorithms used and adjustments to algorithm parameters vary from image to image. Therefore, a "one-size-size-fits-all" approach is not feasible. Optimal image processing requires human assessment by SMEs which makes it time-consuming.

2.4.4.2 Feature Extraction

Feature extraction is usually applied to obtain reliable features for classification and identification of defects in PCBs. Features are regions in an image which help to reduce redundant or noisy information. Features are extracted by either traditional human-developed algorithms or artificial neural networks. Common feature extraction algorithms include, but are not limited to, scale-invariant feature transform (SIFT) [42], speeded up robust features (SURF) [9], corner detection [28], edge detection [13], contour extraction, and RGB/HSV color space information. The selection of features is highly application-driven; it depends on the imaging modality and the target object.

Challenges in Feature Extraction Feature extraction has disadvantages. It is dependent on images acquired by the chosen imaging modality. It is also necessary to have domain knowledge to extract relevant features to detect defects in PCBs. Feature representations that are suitable for the classification of one type of component may not be sufficient for the classification of another. Finally, it may be impossible to choose a feature representation that entirely separates objects into their distinct classes based solely on images taken using a single modality.

2.4.4.3 Classification

The next stage is the classification of components, where resistors, capacitors, LEDs, ICs, and pins will be grouped into different classes. Classification facilitates the defect detection step, where each class could be compared with a golden sample. We discuss three methods used to address classification problems. These methods may be useful for recovering a bill of material (BoM) for a PCB.

Recognition of text on a PCB and on components is vital for BoM extraction [41, 54]. Counterfeit parts can be detected by cross-referencing component serial numbers with OEM data. Furthermore, malicious addition and removal of components can be identified by comparing a detected BoM with the expected BoM. For a more detailed analysis, the texture of the components can be compared with golden samples. For example, the logo on a suspect component could be compared to an ideal component's logo to detect counterfeiting. However, there are many cases where the golden sample is not available. Hence other assurance methods should be used.

Conventional Approaches vs. Learning-Based Approaches Traditionally, classification was performed by comparing patterns and features between two images. Methods such as template matching were used for direct image-to-image comparison, whereas feature matching was performed using correlation and distance metrics. However, with the advent of deep learning, systems capable of simultaneously performing both feature extraction and classification have emerged. Such methods are efficient and can outperform human-selected features. Due to the following reasons, their role in image processing and computer vision has become widespread.

1. **Template Matching:** As the name suggests, template matching classifies image pixels by matching them with known patterns. The matching procedure is usually achieved by correlation-based methods, such as the eigen templates [52], mean averaged templates [7], etc., or the minimization algorithms [68]. However, since template matching may use large groups of pixels to do correlation calculation, it is very time-consuming and is influenced by rotation, scale, and translation, which limits its usefulness for practical applications.
2. **Classifier-Based Learning** Classifier-based learning approaches are usually applied after feature extraction. Classifiers analyze and evaluate feature vectors from image pixels and make decisions to group them based on various dynamic rules. The most well-known classifier is the support vector machine (SVM) [57] and its variations. These apply kernels to address nonlinear problems [34]. Others like Naive Bayes classifier [49], K-nearest neighbor classifier [35], and decision tree [45] are also available for component classification.
3. **Artificial Neural Networks (ANNs):** Artificial neural networks (ANNs) are powerful tools for image localization and classification. In contrast to human-designed feature extraction, ANNs can extract features automatically. Their performance typically improves when they are trained with more data. Considering that limited data is available for PCB assurance compared to other traditional object detection tasks, transfer learning can be useful for training high-quality models. Zhao et al. [72] used only 500 images to fine-tune an existing component recognition network and successfully classified images into 20 classes. However, it is necessary to conduct comprehensive experiments to verify this solution because of the unique characteristics of PCB images. Another possible solution is to use semi-supervised ANNs. In this approach, only a small number of labeled image samples for training is required. The literature has demonstrated that semi-supervised approaches can improve performance in object detection [12, 23, 33].

Challenges in Classification Challenges arise in component classification due to limited data [36] and misrepresentation of extracted features. These challenges may lead to *overfitting*. Overfit models are those that learn patterns unique to their training set which do not generalize well to the set of examples that the model will be used to classify.

2.4.5   Identification

Identification is similar to classification but is used to make more fine-grained distinctions. It recognizes logos, names, and values on components or a PCB. These tasks are necessary for generating a BoM as well as for defect localization in PCBs. Logo identification can be achieved using image matching methods described above. The name and value of components are often printed on components and can thus be acquired by optical character recognition (OCR) systems [41].

There are many OCR engines, each with unique advantages and disadvantages. One of the most widely used engines is called Tesseract, which was developed by HP and is now maintained by Google. It is easy to access and implement with various algorithms implemented [54]. However, the limitations of using the existing OCR system to generate a BoM are apparent. Text on PCBs has unique characteristics. For example, fonts are usually small and may be rotated, and text on ICs may be blurry or low contrast. Moreover, traces and vias on PCB boards may appear similar to characters. Existing OCR systems are mainly built to recognize characters from books or road signs, and most of them are thus not capable of addressing the unique challenges of fonts on PCBs. Therefore, it is necessary to construct an OCR system that meets the needs of PCB assurance.

2.4.5.1   Authentication

After image acquisition and analysis, the final step is to perform authentication. The objective here is to present the results from image analysis in a readable format that can be analyzed by a computer or SME. There are two possible scenarios for authentication:

1. **Scenario 1:** when there is a golden sample, bill of materials, layout, or CAD file available for cross-verification (this is usually the case for OEMs, CEMs, and partner companies with access to product IP and design specifications).
2. **Scenario 2:** when there is no such golden sample or data for cross-verification (this is the case for end users such as individuals or consumer companies). It is essential to understand two different cases within PCB verification, as this will help inform which imaging modalities and imaging algorithms will be used, along with their benefits and limitations. While there is a need to authenticate PCBs in both cases, the accuracy and thoroughness of the inspection will vary.

The process of authentication can be separated into the storage phase and the authentication phase. After image analysis, there is a need to store the obtained information. In the case of PCBs and microelectronics, there are two common options:

1. **Bill of Materials:** The BoM is a spreadsheet or document that lists all components used to fabricate the PCB. The BoM provides the name, location, and placement of each component on the board.
2. **Computer-Aided Design:** CAD is a comprehensive model of the entire PCB. It provides information on the entire PCB layout, including traces and vias from all layers, in addition to the location of all the components.

Both BoM and CAD are widely accepted in the electronic community.

In addition to the information format, the processing of the information for authentication is also application-dependent. As previously mentioned, there are two possible scenarios based on the user type. OEMs and CEMs have access to the golden sample information such as a BoM, CAD files, or even a golden PCB. On the other hand, consumers, end users, and independent groups may not have this golden sample information for comparison. The following two scenarios are described below:

1. **OEM Scenario—Presence of Golden Sample Information:** Golden sample information such as BoMs and CAD models are available to cross-verify the data obtained from a device under test (DUT). If no BoM is present, images of golden samples can also be used directly for matching, or the images can be sent through our AutoBoM framework to generate a BoM first and then used for matching.
2. **End-User Scenario—Absence of Golden Sample Information:** When there is no golden sample present, the end users can only use the DUT PCB to perform authentication. The DUT is sent through the three steps of our AutoBoM framework, and the details obtained are formatted into a BoM or a CAD. This can be stored and sent to an SME for analysis or to the OEM company for cross-verification.

The presented framework can be used by various parties, regardless of their access to golden sample information. OEMs and end users must both make use of multi-modal imaging and complementary image analysis techniques to assess the PCB authenticity.

2.5 Counterfeit Detection Using Electromagnetic Fingerprint

Automation in physical inspection with computer vision and image processing can significantly reduce the time and cost required for counterfeit electronic detection. Solutions based on side-channel analysis can be a cheaper alternative. Near-field electromagnetic (EM) information can be used as a side-channel information source to provide design-specific fingerprints. These fingerprints can be applied to detect recycled, remarked, and cloned ICs. In this section, we discuss EMFORCED framework [55, 56], a non-destructive and low-cost IC authentication approach based on EM traces in the chip. Prior knowledge of node technology, design, test vector, and specific software for EM analysis is not required for this framework.

2.5.1 Near-Field Electromagnetic Emission

A time-varying current signal in a wire generates EM wave. The strength of the EM field is directly proportional to the propagating signal's frequency, the amplitude of the signal, and the length of the wire. Since EM emanations in a device can couple to nearby electronic components and systems, several industry-standard solutions have already been developed to attenuate the emanation within a certain distance and reduce the cross-talk with nearby devices [48].

Modern ICs are fabricated with a number of current-carrying interconnect layers, positioned above the active region of the device. The metal lines in the interconnect layers are the major source of the EM field and act as antennas for the EM wave propagation. Therefore, the activity of ICs can be directly correlated with the near-field EM emanation measurement. As each metal line corresponds to a unique EM emission source, the layout of a given design determines the emission profile emanating from a given device under test (DUT). In any digital or mixed-signal IC, the clock distribution network stems the EM emanation due to fast transition time of the clock edges, its periodicity, and high frequency. The IC draws large amounts of current from the power distribution network during the rising/falling edge of the clock. Therefore, the power and the clock distribution network are the most prominent source of EM radiation.

Distinguishing Characteristic Authenticity of an IC can be verified by examining the EM characteristics of the DUT. The charge carrying metal traces in the IC can be considered as an array of point sources. The intensity of the EM emission, E, accumulated over the entire surface, $x.y$, of an IC can be approximated as:

$$E \propto \int_x \int_y \frac{\vec{S}}{4\pi \vec{r}^2} dxdy = \int_x \int_y \frac{I_{(x,y)}^2 Z_{(x,y)}}{a\pi r_{(x,y)}^2} dxdy \qquad (2.1)$$

where I, Z, S, and r represent the current through the wires, complex load, apparent power consumption of the device, and observation distance, respectively. Equation 2.1 is applicable when the electromagnetic wave is traveling through free space. The effect of wave-propagating medium other than free space, e.g., silicon and packaging materials, can be factored in the Eq. 2.1, using the permittivity constant. This relationship gives insights into the EMI occurring due to different input signals and relative measurement position over the die.

Differences between the circuit designs will materialize as differences in the observed fingerprints. An example is the variations associated with the clock distribution network for different layouts. During clock tree synthesis, different computer-aided design (CAD) tools create significantly different routing networks and buffer locations to minimize the delay, wire length, and skew. Therefore, the EM emission from clock trees of two different designs is expected to be different and acts as an inter-design variable. However, a low intra-design variable is also required for a reliable "fingerprint" generation. This requires two devices containing the same layout to produce a consistent signature within a certain margin. The difference between the EM fingerprint due to intra-design and inter-design variation can be used to detect recycled or cloned IC.

2.5.2   Threat Model

Remarked and cloned counterfeit ICs impose a substantial threat to the stakeholders within the semiconductor supply chain. To determine the primary entities that remarked and cloned ICs have infiltrated, we present the following scenarios:

1. IC design assets, e.g., GDSII and fabrication node, are available to the entity. The entity has access to verified IC to identify the remarked or cloned ICs. Fabless design house and government are examples of such entities.
2. The entity does not have access to the design and a reference signature from a trusted party but needs to replace a legacy component that they possess. Such scenarios are suitable for government and critical system integrators.
3. The entity has acquired the ICs, which match with the EM signature available from a trusted party. The authenticity of the ICs needs to be verified. An example of such an entity is OCM or component retailers.

All the entities mentioned above can scrutinize the authenticity of suspected ICs using EMFORCED framework. Fabless design houses fabricate their designs in offshore foundries. Since the "golden" IC is sourced from the untrusted foundry, verifying the authenticity of the IC is a challenging task. The design house can reverse engineer the entire chip, extract the entire netlist of the chip, and compare the extracted design with "golden" layout or GDSII file available. Though there has been a significant advancement in the automation in delayering, imaging, and the netlist extraction process used in reverse engineering, the entire workflow is still a time-consuming approach and requires subject matter expert (SME) involvement [11]. Approaches like Trojan scanner [62, 67] can significantly reduce the time required for authentication. In the Trojan scanner, the backside of the chip is thinned down to 1 3 μm. The active layer of the IC is imaged with SEM compared with "golden" layout for malicious modification in the IC [66]. However, if the EM fingerprints of the ICs are extracted prior to reverse engineering, the remaining ICs can be confidently categorized depending upon the correlation between their fingerprints and the verified golden IC. Unlike the fabless design house, the component retailers have access to verified reference EM signature, which can be used to identify the counterfeit chips. The entities with access to

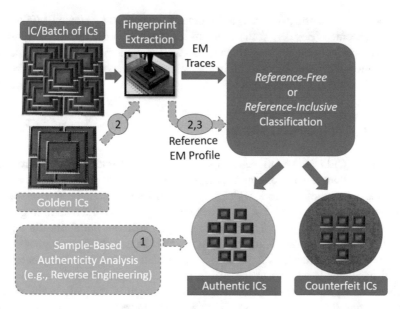

Fig. 2.5 EMFORCED framework overview with scenario-specific elements labeled in orange [55]

legacy components, such as government, can use those components as a source of golden fingerprint and use EMFORCED framework to authenticate the newly acquired suspected devices.

2.5.3   EMFORCED Framework

Figure 2.5 shows the EMFORCED framework used to classify the device under test (DUT) into authentic and counterfeit groupings. An entity can acquire the IC from the open market, distributor (e.g., government, end user), or offshore foundry (e.g., fabless design house). The ICs undergo EM fingerprint acquisition. Machine learning methods are used to process the collected fingerprint and separate the ICs into separate groups. A single sample from each group is analyzed using a suitable approach, e.g., reverse engineering and physical inspection [24], to identify groups of authentic and counterfeit ICs.

2.5.3.1   Scenario Identification

The EMFORCED framework is applicable to the scenarios 1–3 described in Sect. 2.5.2. Therefore, the first step of authenticity verification of suspected ICs is to identify the scenario that best fits the entity interested in counterfeit detection. The scenario corresponding to a given entity impacts the process by which EMFORCED must be executed. Besides, scenario identification is also used in the selection of the reference-free and reference-inclusive classification methods. Reference-free classification and sample-based authenticity analysis methods are applicable for scenario 1, while the others are not.

2.5.3.2   EM Fingerprint Extraction

The EMFORCED framework relies upon near-field EM traces obtained from ICs operating without test vector application or prior knowledge of the IC. Stimulating the clock distribution network with an input signal can generate a device-specific EM signature. The reason for the selection of clock signal for EM fingerprint extraction is twofold: – (a) The clock signal permeates in the entire die and (b) naturally supports an oscillating or pulsed signal. Hence, it allows higher signal-to-noise EM measurements. A pulsed signal present from a function generator or crystal oscillator is applied to the ICs in a loop. The generated EM emission is collected using an industry-standard EM probe, which is available in the market. Implementing the circuit in a custom-made PCB can reduce the time required for EM measurement. Once the chip is situated into the test platform, i.e., custom-made PCB for measurement, the EM probe is centered over the die and lowered until it contacts the surface of the package. The EM traces collected from the near-field probe are amplified and then sampled to generate a digital representation of the device-specific EM fingerprint. The signal from the near-field probe is amplified and then sampled to provide a digital representation of the EM fingerprint. The extracted fingerprints can then be processed and classified into their respective groups.

2.5.3.3   Suspect Classification Methods

Different techniques can be used to identify ICs with similar fingerprints and categorize those ICs into groups. Here, reference-free and reference-inclusive classification methods are described (see Sect. 2.5.3.1 for detail).

Reference-Free Classification

If the entity does not have access to a reference profile, the first step in EMFORCED framework is to separate the suspected chips into groups based on EM signature similarity. If the reference EM profile is not available, the ICs can be separated quickly and efficiently into their respective groups using unsupervised classification methods. Principal component analysis (PCA) can be used for dimensionality reduction and feature extraction [53]. The basis behind PCA is to convert poten-tially correlated variables into a set that is linearly uncorrelated. These correlated variables, termed as principal components, represent the data in a manner which highlights the most expressive features of each signal by projecting the data in orthogonal directions that contain the most variance. Therefore, most variance in data is represented by the first principal component, the second most variance in data is represented by the second principal component, and so forth. Furthermore, the number of distinct principal components in a data set will be defined as the smaller of the number of observations or original variables. By determining the principal components that contain the most variance, one is able to quantify the importance of each dimension of the data and provide a reasonable characterization within a reduced dimensional space.

In our analysis, PCA is used to reduce the 250,000 sample-per-fingerprint feature vector down to a dimensionality of N - 1, where N is the number of measurements used for training. Cross-correlation analysis is another method based on statistical analysis for dividing the ICs into their relative groups. This method provides a correlation of each time-series signal comparison in terms of simple percentage. Therefore, in the EMFORCED framework, cross-correlations serve as a similarity metric to verify that the DUT is a member of one of the groups defined by PCA or reference profile.

Reference-Inclusive Classification

The authenticity of the DUT can be determined using a reference-inclusive classification if the entity has access to a reference profile provided by a trusted party. To confidently address all possible scenarios of remarked and cloned IC insertion, a layered approach must be used. The entity can use cross-correlation for device authenticity checking if she has access to a single reference fingerprint. Either cross-correlation or PCA can be applied if the entity has a number of suspect parts to classify. However, clear fingerprint separation using PCA does not guarantee that the devices are a part of different groups. Should all of the fingerprints used for classification belong to a single device category, PCA may still separate them regardless of their likeness [55]. Cross-correlation analysis should be prioritized in reference-inclusive classification to determine whether the EM emission profile aligns well with the available reference. Several established approaches have been utilized in EMFORCED framework to differentiate remarked and cloned ICs from authentic ICs in reference-inclusive technique. For dimensionality reduction, PCA is used on all genuine reference fingerprints. Once the dimensionality of the collected EM fingerprint of suspected ICs is reduced, similar to reference-free approach, observing the known good groups and determining to which group each suspect DUT belongs, similar to our reference-free approach, will not be solved by simply calculating the distance from the suspect to the group. If the DUT does not get classified into one of the dimensional spaces, then the suspected IC is considered as misclassified. Outlier detection is adopted to address misclassification issue. The outlier method of ENFORCED framework defines an outlier as three standard deviations away from the median. When calculating outliers, not all principal components are equally weighted. For example, our first three principal components account for ≈96% of the variance from the projected value. Each principal component value is individually compared as an outlier 1 or inlier 0 against the corresponding authentic references. This binary value is then multiplied by the principal component weight to define the likelihood that a measurement is an outlier. After outlier detection against all good groups, the supervised machine learning algorithm, linear discriminant analysis (LDA), is used [53]. Group label information and feature vectors is utilized in LDA to form boundaries between the groups. It can classify samples into the most closely related group. This machine learning classification method works by maximizing the distance between the mean values of the class and minimizing the scatter within each class. Using the linear transformation and dimensionality reduction of PCA with the nonlinear multiclass classification of LDA provides increased classification accuracy. LDA can provide the final classification for suspects that classify as inliers in two or more groups, as well as verify the classification of your single-class inliers.

2.5.3.4   Sample-Based Authenticity Analysis

Once the DUTs are differentiated into groups, in reference-free classification, a sample from each group is analyzed for authenticity. The entity can choose the suitable analysis approach, e.g., physical inspection or reverse engineering, depending on the capability available to the entity. Once the sample DUTs have been determined to be either authentic or counterfeit, the remainder of the DUTs can be confidently classified within their designation.

2.5.3.5   Case Study: Counterfeit IC Detection for 8051 Microcontroller

EMFORCED framework is used to identify counterfeit 8051-series microcontroller ICs [55, 56]. The DUT microcontrollers manufactured by three different vendors, Atmel, Maxim, and NXP, are acquired from both "trusted" suppliers and gray market sellers.

Results from Reference-Free Analysis of DUT

The DUT verification is performed on two different platforms, (a) breadboard implementation for reference-free classification and (b) COTS 8051 development board for the reference-inclusive classification (see Fig. 2.6). The breadboard setup and development board used on an external clock signal from a function generator and crystal oscillator on the PCB, respectively. The near-field EM radiation is measured from a single direction using a factory-tuned EM probe. An EM shield around the setup is required to suppress the noise from external sources, e.g., PCB wiring, nearby equipment, etc. A mixed-signal oscilloscope is used with a sampling rate of 25 GS/s for all data acquisition. Sampling the clock frequency above the

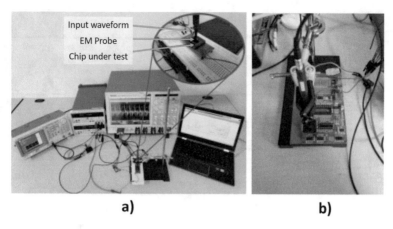

Input waveform
EM Probe
Chip under test

a)   b)

Fig. 2.6 EMFORCED measurement environments: (**a**) breadboard implementation and (**b**) development board implementation [55]

Nyquist frequency enables transient effects to be extracted and used for accurate classification.

1. **Breadboard Implementation:** In breadboard implementation, a 5 V peak-to-peak, 16 MHz square wave with 50% duty cycle is used as an input clock signal. The signal is provided by a dedicated function generator at 16 MHz to comply with operating frequency of the DUT. To ensure that the collected EM signatures are solely dependent upon the device characteristics, the voltage, frequency, and pulse shape remained the same for all the devices. A Langer EMV RF-K 7-4 near-field probe was used for fingerprint extraction with the breadboard implementation. This relatively large (6 mm × 10 mm) probe provides approximately 5 mm of placement accuracy for manual probe alignment.
2. **Development Board Implementation:** The development board provides all required pull-up resistors to boot into an operating state and is readily available to entities as a standard COTS component. A crystal oscillator in the development board acts as the source of clock signal for the DUT.

Upon physical inspection of the suspected ICs from the gray market, components appeared fairly different from the authentic ICs. A couple of Atmel suspected chips would be flagged due to deference in marking from the trusted ones. Additionally, the markings on the Maxim and NXP suspects strongly deviated from their authentic counterparts (see Fig. 2.7). The observations made from physical inspection should not be considered conclusive in determining the authenticity of a given DUT. These variations could be seen even between the original and second Atmel authentic groups. Table 2.3 summarizes the comparison of all 640 fingerprints against one another. Grouping ICs based upon their vendor and date of acquisition, the seven categories that we tested are provided. The average intra-group cross-correlation determines how alike ICs of a given group are to one another. The threshold

Fig. 2.7 Visual inspection comparison of authentic and gray market ICs [55]

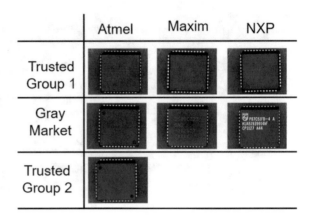

Table 2.3 Cross-correlation analysis comparing intra-group, intra-vendor, and inter-vendor segments of 8051 fingerprint data

Device type	Average intra-group X-corr	Exclusive average intra-vendor X-corr	Average inter-vendor X-corr
Atmel trusted group 1	92.3%	93.3%	49.4%
Atmel trusted group 2	96.9%	94.6%	50.5%
Atmel gray market	96.3%	90.7%	48.7%
Maxim trusted	97.9%	94.9%	55.0%
Maxim gray market	96.4%	94.6%	54.8%
NXP trusted	90.7%	61.4%	53.8%
NXP gray market	88.4%	52.5%	44.1%

for cross-correlation is defined at or above 90%. The value is decided based on empirical analysis. In Table 2.3, all the gray market ICs, except the gray market NXP chips, appear well correlated within their respective groups. The exclusively averaged intra-vendor cross-correlation describes how well the tested group aligns with the other group(s) from the same vendor. It has been identified that the NXP trusted and gray market ICs behave very differently, while both Atmel and Maxim groups are very well correlated with intra-vendor cross-correlations. The average inter-vendor cross-correlation is also provided from another IC, to evaluate whether the NXP DUTs are counterfeit or not. However, still the cross-correlation remains low over all device groups.

Results from Reference-Inclusive Analysis of DUT

In order to identify counterfeit ICs, a PCA model is trained using the reference fingerprints. Figure 2.8 represents the projections of all authentic fingerprints plotted using their first three principal components. Next, the suspect fingerprint measurements from the gray market ICs are projected into our existing PCA model. The projected suspect data (denoted by an S) has been overlaid onto the

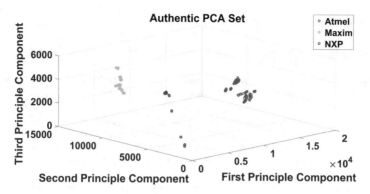

Fig. 2.8 Projected (three-dimensional) PCA training data for all authentic ICDs using the first three principal components [55]

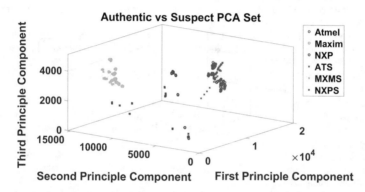

Fig. 2.9 A comparison between authentic and suspect fingerprints using the first three components of their PCA projection. ATS, MXMS, NXPS—refer to suspect ICs of the respective manufacturers [55]

reference data in Fig. 2.9. Figure 2.9 shows that, though DUT from Atmel and Maxim mostly remain within the tolerable authentic region, the NXP suspected ICs deviate significantly. This suspicious behavior reiterates the results acquired in the reference-free cross-correlation analysis.

2.6 Conclusions

Detection and prevention of counterfeit electronic components have become significant challenges in the electronic component supply chain. In this chapter, we presented various physical inspection-based counterfeit detection approaches. We have also presented the opportunities and challenges of image processing and computer vision applications in counterfeit detection. This chapter also presented another emerging counterfeit detection approach based on EM fingerprinting.

References

1. Frame averaging, 2018. http://www.4pi.com/teksupport/Rev10online/manual/39_4-frameavg. htm
2. Reports of Counterfeit Parts Quadruple Since 2009, Challenging US Defense Industry and National Security. https://cdn.ihs.com/www/pdf/investor-relations/IHS-2011-Annual-Report. pdf
3. Ahi, K., Asadizanjani, N., Shahbazmohamadi, S., Tehranipoor, M., Anwar, M.: Terahertz characterization of electronic components and comparison of teraherlz imaging with x-ray imaging techniques. In: Terahertz Physics, Devices, and Systems IX: Advanced Applications in Industry and Defense, vol. 9483, p. 94830K. International Society for Optics and Photonics (2015)
4. Amir, S., Shakya, B., Forte, D., Tehranipoor, M., Bhunia, S.: Comparative analysis of hardware obfuscation for ip protection. In: Proceedings of the on Great Lakes Symposium on VLSI 2017, pp. 363–368 (2017)
5. Asadizanjani, N., Shahbazmohamadi, S., Tehranipoor, M., Forte, D.: Non-destructive pcb reverse engineering using x-ray micro computed tomography. In: 41st International symposium for testing and failure analysis. ASM, pp. 1–5 (2015)
6. Asadizanjani, N., Tehranipoor, M., Forte, D.: PCB reverse engineering using nondestructive x-ray tomography and advanced image processing. IEEE Trans Compon. Packag Manuf. Technol. **7**(2), 292–299 (2017)
7. Ashburner, J., Friston, K.J.: Computing average shaped tissue probability templates. Neuroimage **45**(2), 333–341 (2009)
8. Azhagan, M., Mehta, D., Lu, H., Agrawal, S., Tehranipoor, M., Woodard, D.L., Asadizanjani, N., Chawla, P.: A review on automatic bill of material generation and visual inspection on pcbs. In: ISTFA 2019: Proceedings of the 45th International Symposium for Testing and Failure Analysis, p. 256. ASM International (2019)
9. Bay, H., Tuytelaars, T., Van Gool, L.: Surf: Speeded up robust features. In: A. Leonardis, H. Bischof, A. Pinz (eds.) Computer Vision – ECCV 2006, pp. 404–417. Springer, Berlin/Heidelberg (2006)
10. Baygin, M., Karakose, M., Sarimaden, A., Akin, E.: Machine vision based defect detection approach using image processing. In: 2017 International Artificial Intelligence and Data Processing Symposium (IDAP), pp. 1–5 (2017)
11. Botero, U.J., Wilson, R., Lu, H., Rahman, M.T., Mallaiyan, M.A., Ganji, F., Asadizanjani, N., Tehranipoor, M.M., Woodard, D.L., Forte, D.: Hardware trust and assurance through reverse engineering: A survey and outlook from image analysis and machine learning perspectives. arXiv preprint arXiv:2002.04210 (2020)
12. Camps-Valls, G., Marsheva, T.V.B., Zhou, D.: Semi-supervised graph-based hyperspectral image classification. IEEE Trans. Geosci. Remote Sens. **45**(10), 3044–3054 (2007)
13. Canny, J.: A computational approach to edge detection. In: Readings in Computer Vision, pp. 184–203. Elsevier (1987)
14. Cobb, W.E., Garcia, E.W., Temple, M.A., Baldwin, R.O., Kim, Y.C.: Physical layer identification of embedded devices using rf-dna fingerprinting. In: 2010-Milcom 2010 Military Communications Conference, pp. 2168–2173. IEEE (2010)
15. Committee, S.A.S., et al.: Inquiry into counterfeit electronic parts in the department of defense supply chain. U.S. G.P.O., Washington, DC (2012)
16. Dogan, H., Alam, M.M., Asadizanjani, N., Shahbazmohamadi, S., Forte, D., Tehranipoor, M.: Analyzing the impact of x-ray tomography on the reliability of integrated circuits. In: Proceedings from the 41st International Symposium for Testing and Failure Analysis (ISTFA), pp. 1–10 (2015)

17. Dogan, H., Forte, D., Tehranipoor, M.M.: Aging analysis for recycled fpga detection. In: 2014 IEEE International Symposium on Defect and Fault Tolerance in VLSI and Nanotechnology Systems (DFT), pp. 171–176. IEEE (2014)

18. Dougherty, E.: An introduction to morphological image processing. Tutorial texts in optical engineering. SPIE Optical Engineering Press (1992). https://books.google.com/books?id=1kvxAAAAMAAJ

19. EDN: Guide for spotting counterfeit cisco equipment. https://www.edn.com/guide-for-spotting-counterfeit-cisco-equipment/

20. Friedman, N., Russell, S.: Image segmentation in video sequences: A probabilistic approach. In: Proceedings of the Thirteenth conference on Uncertainty in artificial intelligence, pp. 175–181. Morgan Kaufmann Publishers Inc. (1997)

21. Ghosh, P., Chakraborty, R.S.: Recycled and remarked counterfeit integrated circuit detection by image-processing-based package texture and indent analysis. IEEE Trans. Ind. Inf. **15**(4), 1966–1974 (2018)

22. Ghosh, P., Forte, D., Woodard, D.L., Chakraborty, R.S.: Automated detection of pin defects on counterfeit microelectronics. In: ISTFA 2018: Proceedings from the 44th International Symposium for Testing and Failure Analysis, p. 57. ASM International (2018)

23. Guillaumin, M., Verbeek, J., Schmid, C.: Multimodal semi-supervised learning for image classification. In: 2010 IEEE Computer Society Conference on Computer Vision and Pattern Recognition, pp. 902–909. IEEE (2010)

24. Guin, U., DiMase, D., Tehranipoor, M.: Counterfeit integrated circuits: Detection, avoidance, and the challenges ahead. J. Electron. Test. **30**(1), 9–23 (2014)

25. Guin, U., Forte, D., Tehranipoor, M.: Anti-counterfeit techniques: From design to resign. In: 2013 14th International Workshop on Microprocessor Test and Verification, pp. 89–94. IEEE (2013)

26. Guin, U., Forte, D., Tehranipoor, M.: Design of accurate low-cost on-chip structures for protecting integrated circuits against recycling. IEEE Trans. Very Large Scale Integr. VLSI Syst. **24**(4), 1233–1246 (2015)

27. Guin, U., Zhang, X., Forte, D., Tehranipoor, M.: Low-cost on-chip structures for combating die and ic recycling. In: 2014 51st ACM/EDAC/IEEE Design Automation Conference (DAC), pp. 1–6. IEEE (2014)

28. Harris, C.G., Stephens, M., et al.: A combined corner and edge detector. In: Alvey Vision Conference, vol. 15, pp. 10–5244. Citeseer (1988)

29. Hsieh, J.: Adaptive streak artifact reduction in computed tomography resulting from excessive x-ray photon noise. Med. Phys. **25**(11), 2139–2147 (1998)

30. Hwang, H., Haddad, R.A.: Adaptive median filters: new algorithms and results. IEEE Trans. Image Process. **4**(4), 499–502 (1995)

31. Ito, K.: Gaussian filter for nonlinear filtering problems. In: Proceedings of the 39th IEEE Conference on Decision and Control (Cat. No. 00CH37187), vol. 2, pp. 1218–1223. IEEE (2000)

32. Jain, A.K.: Fundamentals of digital image processing. Prentice Hall, Englewood Cliffs (1989)

33. Kingma, D.P., Mohamed, S., Rezende, D.J., Welling, M.: Semi-supervised learning with deep generative models. In: Advances in Neural Information Processing Systems, pp. 3581–3589. Arvix.org (2014)

34. Levenberg, K.: A method for the solution of certain non-linear problems in least squares. Q. Appl. Math. **2**(2), 164–168 (1944)

35. Liao, Y., Vemuri, V.R.: Use of k-nearest neighbor classifier for intrusion detection. Comput. Secur. **21**(5), 439–448 (2002)

36. Lu, H., Mehta, D., Paradis, O., Asadizanjani, N., Tehranipoor, M., Woodard, D.L.: Fics-pcb: A multi-modal image dataset for automated printed circuit board visual inspection

37. Malge, P., Nadaf, R.: Pcb defect detection, classification and localization using mathematical morphology and image processing tools. Int. J. Comput. Appl. **87**(9) (2014)

38. Mehta, D., Lu, H., and Paradis, O.P., MS, M.A., Rahman, M.T., Iskander, Y., Chawla, P., Woodard, D.L., Tehranipoor, M., Asadizanjani, N.: The big hack explained: Detection and prevention of PCB supply chain implants. ACM J. Emerg. Technol. Comput. Syst. (JETC) **16**(4), 1–25 (2020)
39. Miller, M., Meraglia, J., Hayward, J.: Traceability in the age of globalization: a proposal for a marking protocol to assure authenticity of electronic parts. Technical report, SAE Technical Paper (2012)
40. Moganti, M., Ercal, F., Dagli, C.H., Tsunekawa, S.: Automatic pcb inspection algorithms: a survey. Comput. Vis. Image Underst. **63**(2), 287–313 (1996)
41. Mori, S., Nishida, H., Yamada, H.: Optical character recognition. Wiley Inc., New York (1999)
42. Ng, P.C., Henikoff, S.: Sift: Predicting amino acid changes that affect protein function. Nucleic Acids Res. **31**(13), 3812–3814 (2003)
43. Nikam, P.A., Sawant, S.D.: Circuit board defect detection using image processing and microcontroller. In: 2017 International Conference on Intelligent Computing and Control Systems (ICICCS), pp. 1096–1098 (2017)
44. Piccardi, M.: Background subtraction techniques: a review. In: 2004 IEEE International Conference on Systems, Man and Cybernetics (IEEE Cat. No.04CH37583), vol. 4, pp. 3099–3104 (2004). https://doi.org/10.1109/ICSMC.2004.1400815
45. Quinlan, J.R.: Induction of decision trees. Mach. Learn. **1**(1), 81–106 (1986)
46. Rahman, M.T., Rahman, M.S., Wang, H., Tajik, S., Khalil, W., Farahmandi, F., Forte, D., Asadizanjani, N., Tehranipoor, M.: Defense-in-depth: a recipe for logic locking to prevail. Integration **72**, 39–57 (2020)
47. Rahman, M.T., Shi, Q., Tajik, S., Shen, H., Woodard, D.L., Tehranipoor, M., Asadizanjani, N.: Physical inspection & attacks: new frontier in hardware security. In: 2018 IEEE 3rd International Verification and Security Workshop (IVSW), pp. 93–102. IEEE (2018)
48. Ramdani, M., Sicard, E., Boyer, A., Dhia, S.B., Whalen, J.J., Hubing, T.H., Coenen, M., Wada, O.: The electromagnetic compatibility of integrated circuits—past, present, and future. IEEE Trans. Electromagn. Compat. **51**(1), 78–100 (2009)
49. Rish, I., et al.: An empirical study of the naive bayes classifier. In: IJCAI 2001 Workshop on Empirical Methods in Artificial Intelligence, vol. 3, pp. 41–46 (2001)
50. Savitzky, A., Golay, M.J.: Smoothing and differentiation of data by simplified least squares procedures. Anal. Chem. **36**(8), 1627–1639 (1964)
51. Shahbazmohamadi, S., Forte, D., Tehranipoor, M.: Advanced physical inspection methods for counterfeit ic detection. In: ISTFA 2014: Conference Proceedings from the 40th International Symposium for Testing and Failure Analysis, p. 55. ASM International (2014)
52. Shaw, A.K., Bhatnagar, V.: Automatic target recognition using eigen templates. In: Algorithms for Synthetic Aperture Radar Imagery V, vol. 3370, pp. 448–460. International Society for Optics and Photonics (1998)
53. Shlens, J.: A tutorial on principal component analysis. arXiv preprint :1404.1100 (2014)
54. Smith, R.: An overview of the tesseract ocr engine. In: Ninth International Conference on Document Analysis and Recognition (ICDAR 2007), vol. 2, pp. 629–633. IEEE (2007)
55. Stern, A., Botero, U., Rahman, F., Forte, D., Tehranipoor, M.: Emforced: Em-based fingerprinting framework for remarked and cloned counterfeit ic detection using machine learning classification. IEEE Transactions on Very Large Scale Integration (VLSI) Systems (2019)
56. Stern, A., Botero, U., Shakya, B., Shen, H., Forte, D., Tehranipoor, M.: Emforced: Em-based fingerprinting framework for counterfeit detection with demonstration on remarked and cloned ics. In: 2018 IEEE International Test Conference (ITC), pp. 1–9. IEEE (2018)
57. Suykens, J.A., Vandewalle, J.: Least squares support vector machine classifiers. Neural. Process. Lett. **9**(3), 293–300 (1999)
58. Tehranipoor, M., Koushanfar, F.: A survey of hardware trojan taxonomy and detection. IEEE Des. Test Comput. **27**(1), 10–25 (2010)
59. Tehranipoor, M., Salmani, H., Zhang, X.: Integrated Circuit Authentication, vol. 10, pp. 978–973. Springer, Cham (2014)

60. Tehranipoor, M., Wang, C.: Introduction to Hardware Security and Trust. Springer Science & Business Media (2011)
61. Tehranipoor, M.M., Guin, U., Forte, D.: Counterfeit integrated circuits. In: Counterfeit Integrated Circuits, pp. 15–36. Springer (2015)
62. Tehranipoor, M.M., Shen, H., Vashistha, N., Asadizanjani, N., Rahman, M.T., Woodard, D.: Hardware trojan scanner (2020). US Patent App. 16/573,922
63. The FBI: Florida Man Charged in Federal Counterfeit Case for Trafficking Bogus Automotive Devices 'Reverse Engineered' in China. Technical report
64. The United States Department of Justice : Departments of Justice and Homeland Security Announce 30 Convictions, More Than $143 Million in Seizures from Initiative Targeting Traffickers in Counterfeit Network Hardware. Technical report
65. Torrance, R., James, D.: The State-of-the-Art in IC Reverse Engineering. In: Proceedings of the International Workshop on Cryptographic Hardware and Embedded Systems (CHES), pp. 363–381 (2009)
66. Vashistha, N., Lu, H., Shi, Q., Rahman, M.T., Shen, H., Woodard, D.L., Asadizanjani, N., Tehranipoor, M.: Trojan scanner: Detecting hardware trojans with rapid sem imaging combined with image processing and machine learning. In: ISTFA 2018: Proceedings from the 44th International Symposium for Testing and Failure Analysis, p. 256. ASM International (2018)
67. Vashistha, N., Rahman, M.T., Shen, H., Woodard, D.L., Asadizanjani, N., Tehranipoor, M.: Detecting hardware trojans inserted by untrusted foundry using physical inspection and advanced image processing. J. Hardw. Syst. Secur. 2(4), 333–344 (2018)
68. Wang, D.Z., Wu, C.H., Ip, A., Chan, C.Y., Wang, D.W.: Fast multi-template matching using a particle swarm optimization algorithm for pcb inspection. In: M. Giacobini, A. Brabazon, S. Cagnoni, G.A. Di Caro, R. Drechsler, A. Ekárt, A.I. Esparcia-Alcázar, M. Farooq, A. Fink, J. McCormack, M. O'Neill, J. Romero, F. Rothlauf, G. Squillero, A.Ş. Uyar, S. Yang (eds.) Applications of Evolutionary Computing, pp. 365–370. Springer, Berlin/Heidelberg (2008)
69. Wu, W.Y., Wang, M.J.J., Liu, C.M.: Automated inspection of printed circuit boards through machine vision. Comput. Ind. 28(2), 103–111 (1996)
70. Xiao, K., Forte, D., Jin, Y., Karri, R., Bhunia, S., Tehranipoor, M.: Hardware trojans: lessons learned after one decade of research. ACM Trans. Des. Autom. Electron. Syst. (TODAES) 22(1), 1–23 (2016)
71. Zhang, X., Tehranipoor, M.: Design of on-chip lightweight sensors for effective detection of recycled ics. IEEE Trans. Very Large Scale Integr. VLSI Syst. 22(5), 1016–1029 (2013)
72. Zhao, W.: Research on the deep learning of the small sample data based on transfer learning. In: AIP Conference Proceedings, vol. 1864, p. 020018. AIP Publishing (2017)

Chapter 3
Physical Inspection of Integrated Circuits

3.1 Introduction

Often, software programmers overlook the security vulnerabilities of hardware systems, assuming the physical components are trusted by default. Cybersecurity engineers and software developers never doubt the integrity and authenticity of the hardware platforms (integrated circuit logic, printed circuit boards); hence, it is assumed that these platforms are reliable, secure, and trustworthy. Due to this widespread belief, adversaries have historically focused their system-level attacks on compromising software rather than the hardware it runs on. Software can be attacked by injecting malware (virus, Trojans), phishing, SQL injection, or denial of service (DoS). To combat these approaches, hardware devices are used to protect the software (root of trust) from various forms of piracy and unauthorized access. Secure hardware tokens are used for storing cryptography keys and generating one-time-use random passwords for two-step authentication. Eventually, these practices reduced surfaces for software attacks. However, this rigor did not extend to hardware exploits. As such, attackers eventually shifted their focus to this realm of physical system exploits. Due to this increase in the number of hardware attacks, it has become essential to pay attention to hardware security. There is also a need for *hardware assurance*, which can be broadly defined as the practice of making hardware secure or ways to validate hardware before it is used. There are two broad techniques to ensure hardware assurance for ICs such as electrical testing (logic testing, side-channel analysis) and physical inspection (full-blown or partial reverse engineering) by performing nano-imaging analysis [7, 24]. In this chapter, we will explore these physical inspection techniques in detail.

An electronic hardware system is composed of one or more PCBs (printed circuit boards) mounted with different hardware components called integrated circuits or semiconductor chips. Each chip is composed of an array of millions to billions of transistors, diodes, resistors, and capacitors as the building blocks of these integrated

N. Asadizanjani et al., *Physical Assurance*, https://doi.org/10.1007/978-3-030-62609-9_3

circuits. These integrated circuits are fabricated on a single piece of silicon called a silicon wafer or a substrate. An IC can be a single module such as a power supply, amplifier, ADC/DAC, microprocessor, or memory. It could also be a complex design such as a system-on-chip (SoC) with all of the aforementioned modules implemented on a single chip. If these ICs are designed for a specific application or targeted product such as a communication system, enterprise server, or automobile, they are called application-specific integrated circuits (ASIC). There is another segment of ICs that can be programmed by an end user after IC manufacturing is complete. These are called field-programmable gate arrays (FPGAs). All ICs enter into the market after a long process of design, fabrication, and testing as defined in the supply chain.

3.2 IC Supply Chain

The IC supply chain (see Fig. 3.1) begins with customer specifications or market demand for a particular product. After defining specifications, these ICs are designed using CAD (computer-aided design) tools which facilitate the integration of a new design with existing IP (intellectual property). These designs are converted into a floor plan called a *layout*, which guides the fabricated structures. After verifying specification and other essential parameters, such as timing, power, and design rules, the final output GDSII file is sent to a foundry.

Many copies of a circuit are fabricated at once on silicon wafers. Each circuit is in the shape of a well-defined square or rectangular area. Each area is referred to as a *die*. After fabrication, these dies are tested on the wafer to validate the design specifications. If a die does not meet certain criteria, it is marked as a failure. Passing dies are cut from the wafer and packaged. These packaged ICs are tested again before advancing to the assembly stage. Among all the previously mentioned steps, the fabrication step is most expensive due to an initial investment of billions of dollars in operation and maintenance of an ultra-clean and precisely controlled fabrication space for manufacturing. Moreover, fabrication facilities require continuous investment in new equipment to keep up with advanced technology nodes. IC design houses have started outsourcing fabrication to overseas foundries to keep the cost of semiconductor fabrication low, reduce their time to market, and, hence, minimize the cost of their chips.

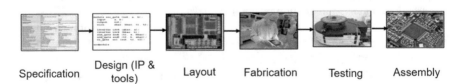

Specification   Design (IP &   Layout   Fabrication   Testing   Assembly
                    tools)

Fig. 3.1 IC supply chain cycle

3.3   Trust and Security Issues

As discussed earlier, a software design loophole can be vulnerable and exploited by adversaries for piracy and cyber attacks. Similarly, the hardware circuits can be attacked by any entity in the supply chain.

3.3.1   Attack Model

Among many entities involved in the supply chain, the primary concern is the overseas fabrication facilities. Outsourcing of semiconductor fabrication has introduced trust issues between design house and foundry, because the latter has full access to all the design details, including GDSII layout, netlist, and test vectors. Also, these foundries have a highly confidential fabrication process which is not shared with the IP owner. Hence, such facilities represent untrusted links in the supply chain. These foundries can overproduce chips (IC/IP piracy), sell out-of-specification ICs on the gray market, and can make malicious modifications (hardware Trojans) to an IC design to undermine its security and performance [28]. The untrusted fabrication facility is one of the most dangerous and discussed threat models [31] in the hardware security community (see Fig. 3.2).

3.3.2   Security Threats in IC Manufacturing

Untrusted supply chain entities pose many threats including, but not limited to:

1. **Hardware Trojans:** A hardware Trojan is a malicious insertion, modification, or deletion of logic cells/transistors. A hardware Trojan could control the whole or a part of a system, cause data leakage, or result in early system failure. For example, data servers and financial institutions are attractive targets for attackers to steal secret information, which may lead to financial loss. Government and military systems are also likely targets for Trojan attacks because they could hamstring combat capabilities or disable critical infrastructure services [31].

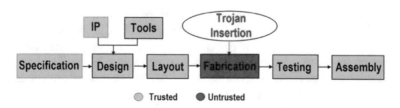

Fig. 3.2 Untrusted foundry attack model

In the original circuit, a hardware Trojan can be introduced at any stage of the design cycle and supply chain such as IP integration, IC design, and offshore fabrication or even at system level by embedding malicious components on printed circuit board (PCB). Among all possible adversaries, offshore foundries are the most threatening as they have full control over the fabrication process and complete access to IC design layout, i.e., GDSII, and test vectors/responses. Therefore, it is quite challenging to prevent hardware Trojan insertion by an offshore foundry during the fabrication stage.

2. **Reverse Engineering:** Reverse engineering is used to steal the design information from an IC by destructive physical inspection. These attacks are performed by de-packaging a chip to take out the silicon die by using strong solvents such as fuming nitric acid, chemical or mechanical polishing, or plasma/laser-based cutting. The exposed die is further polished to expose device layers followed by nano-imaging techniques such as scanning electron microscope (SEM), transmission electron microscope (TEM), or helium ion microscope. This process can be performed from the frontside or backside of the die until all the device layers have been imaged [7, 25]. These images can be further analyzed to reconstruct the circuit design to steal intellectual property (IP). Recent automation in IC deprocessing and imaging techniques and netlist extraction have reduced the operator involvement in the entire reverse engineering process [25], thus reducing the time and cost of reverse engineering.

3. **Counterfeiting:** Counterfeiting of ICs entails either overproduction, releasing out-of-specification parts, changes in the grade of ICs (e.g., changing an automotive/commercial-grade IC to military-grade IC by remarking the package), or selling recycled ICs by scrapping them from recycled/trashed electronic items. Counterfeits generally look similar enough to the original IC that they could be mistaken as genuine by vendors and end users.

Counterfeiting is often performed by untrusted foundries or unscrupulous entities who purchase out-of-specification ICs from foundries and electronic waste recycling units. These ICs enter into the supply chain through distributors or unauthorized selling agents and are sold as genuine products.

3.4 Steps for Hardware Assurance

Methods to ensure hardware trust and assurance can be broadly classified into electrical testing and physical inspection. Electrical testing measures electrical parameters and side channels and searches for anomalies. A rise in temperature, increased resource utilization, power, current consumption, and EM emissions could all be useful for detecting malicious alterations to a circuit. PUF (challenge-response pairs) can be used to detect IP/IC piracy. IDDQ/IDDT (quiescent/transient current) tests can measure indications of aging on ICs to detect recycled components. However, electrical tests have limitations. Circuit noise, insufficient quantity of test patterns for triggering hardware Trojans, and reliance on a golden model all pose practical challenges to electrical testing.

Another method is physical inspection of ICs by nano-level imaging to detect hardware Trojans. In the past, little attention has been paid to physical inspection techniques due to the low accuracy of the polishing methods and the high cost of advanced imaging instruments required for these techniques. However, with advances in SEM, the new instruments can image at a resolution of tens of nanometer, and helium ion microscopy can further image at the sub-nanometer level. These images can be validated by comparing them with a layout designed in a trusted design house for trust validation and hardware assurance.

3.4.1   Physical Inspection Methods

The physical inspection method involves data acquisition by imaging, preprocessing data to make it suitable for feature extraction, and classification. Physical inspection methods that involve sophisticated imaging tools, such as SEM, need an extra preliminary step of sample preparation.

3.4.2   Sample Preparation

Packaged ICs are required to be decapsulated to expose the silicon die. Mechanical polishing is one of the most commonly used techniques to remove the packaging material. However, even after the die is exposed, it is still not ready for SEM imaging because the electrons cannot penetrate a thick layer of the silicon substrate. The substrate needs to be further thinned by using precise polishing methods. Also, the bare die used in the technique is not flat, and curvature keeps changing during the polishing; this change may cause uneven silicon substrate removal over the chip. To mitigate these issues, advanced mechanical polishing techniques such as MultiPrep and VarioMill can be used to perform backside thinning up to 1–2 μm. Using a 5-axis computer numerical control tool combined with an interferometer to adapt the polishing rate and the curvature shape is tracked to ensure uniform thinning across the die.

A smart card die (see Fig. 3.3) is encapsulated into a thin epoxy resin packaged into a plastic shield on one side and a metallic contact pad on the other side. Smart card chip decapsulation begins with removing the die by cutting the package with a sharp cutter. The die, which is covered by epoxy resin, can be further decapsulated using a few drops of fuming nitric acid followed by acetone and isopropyl alcohol wash. Finally, the bare die can be backside thinned by using mechanical polishing.

Fig. 3.3 Sample preparation:
(**a**) smart card, (**b**) die
removal, (**c**) bare die, and (**d**)
backside thinned die [30]

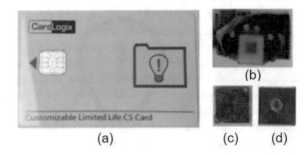

(b)

(a)   (c)   (d)

3.4.3 Data Acquisition: Imaging

The objective here is to take SEM images of the whole die in a short time while
capturing enough details at the active layer for hardware Trojan detection or of
multiple layers for full-blown reverse engineering. Hence, capturing SEM images
with a proper resolution is an essential step in physical inspection.

3.4.3.1 Imaging Parameters

The quality of the SEM images depends on the following SEM parameters. The
effects of these parameters on imaging depend upon the following parameters and
can be clearly observed from Fig. 3.4.

1. **Beam Voltage:** The accelerating voltage (in kV) of the electron beam decides
   the penetration depth of electrons inside the object. For example, a 5 kV beam
   can expose active regions while imaging from the backside, whereas 10 kV can
   further expose subsurface features, including the polysilicon and higher metal
   layers.
2. **Field of View (FOV):** FOV is the area covered by SEM in a single raster. FOV is
   inversely related to the magnification of the image. A large field of view covers
   more features but may lead to blurred images due to low magnification. Imaging
   time increases with a decrease in the field of view.
3. **Dwelling Time (Speed):** Dwelling time is the duration over which the SEM
   detector integrates signals to image one pixel. A higher dwelling time increases
   the signal-to-noise ratio of the image and gives better-quality images but
   increases imaging time. Longer dwelling times also affect surface charge, which
   can introduce imaging artifacts.
4. **Resolution:** Resolution denotes the pixel counts in the image. A higher-
   resolution image is more clear and sharp, but it takes much more time to capture
   high-resolution images.

After setting the abovementioned parameters, the microscope can be programmed
to scan a die in by imaging small areas of the die and stitching the many windows
together to create a complete image.

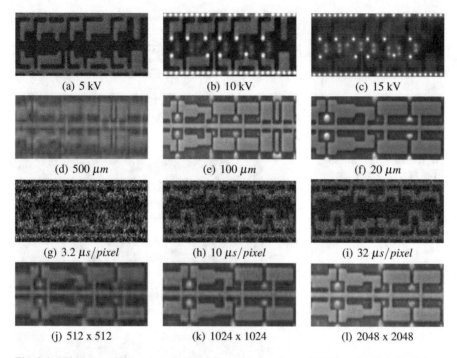

(a) 5 kV   (b) 10 kV   (c) 15 kV

(d) 500 μm   (e) 100 μm   (f) 20 μm

(g) 3.2 μs/pixel   (h) 10 μs/pixel   (i) 32 μs/pixel

(j) 512 x 512   (k) 1024 x 1024   (l) 2048 x 2048

Fig. 3.4 SEM images [30] variations with different beam voltages [(**a**),(**b**), and (**c**)], field of views [(**d**),(**e**), and (**f**)], dwelling times [(**g**),(**h**), and (**i**)], and resolutions [(**j**),(**k**), and (**l**)]

3.4.3.2   Imaging Time

The SEM imaging data in Table 3.1 summarizes the SEM image acquisition time to finish scanning of a 1.5 mm × 1.5 mm die, with different fields of view and dwelling times for 2048 × 2048 resolution. Figure 3.4 and the data in Table 3.1 show that the images captured with a large field of view and small dwelling time take less imaging time, but are unsuitable for detecting changes. A small field of view with long dwelling time captures superior images but collects more data than required which takes an unacceptable amount of time. There is a trade-off between imaging time and the suitability of images for Trojan detection. Imaging time can be carefully chosen based on feedback from image analysis performed in the next step.

3.4.4   Image Preprocessing

A raw SEM image is unsuitable for direct application of image analysis algorithms to detect changes. Hence, a couple of issues need to be addressed first by using

Table 3.1 SEM imaging time variation over dwelling time and field of view [29]

Dwelling time (μs/pixel)	Field of view			
	1500	500	100	20
1	6 s	54 s	22 min 30 s	9 h 23 min
3.2	14 s	2 min 5 s	52 min 5 s	21 h 42 min
10	1 min 25 s	6 min 25 s	5 h 19 min	132 h 49 min
32	2 min 52 s	24 min	10 h 45 min	265 h 30 min

Fig. 3.5 Image preprocessing. (**a**) Original SEM image, (**b**) histogram equalization, (**c**) Gaussian filtering, (**d**) median filtering, and (**e**) thresholding

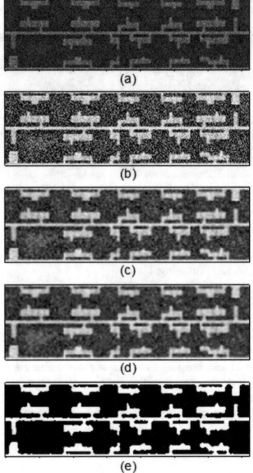

image preprocessing techniques such as alignment of tiled images with each other and enhancement of image features (noise removal, intensity correction, segmentation) for better feature detection.

1. **Histogram Equalization:** This method enhances image contrast by flattening pixel intensities so that the intensity of features can be better distributed on the histogram. This increases the contrast of SEM image features, such as logic gates and the doping regions, against the background and makes them more clearly distinguishable. A drawback of histogram equalization is that it may increase noise, but noise can be reduced using other Gaussian blur and median filtering techniques. Histogram equalization is well-established in applications using medical and radar image processing [13].

2. **Gaussian Blur:** Gaussian white noise is considered as the predominant type of noise in the SEM image. A low-pass filter called a Gaussian filter with an optimized kernel size can be applied to address this problem.

3. **Median Filter:** Median filtering is a smoothening technique for noise removal that preserves edges. Preserving edges is crucial, as edge detection accuracy is necessary for identifying footprints of logic gates. The features in SEM images are sometimes very thin (five to ten pixels), so a small (radius of 3) median filter should be used.

4. **Thresholding:** Thresholding converts grayscale images into binary images. It can remove dark backgrounds and segment active region features for generating a shape descriptor. The intensity value of the cell's active region may not be uniformly distributed within the SEM image, leading to segmentation inaccuracy if standard binary thresholding is used. A more sophisticated algorithm, such as Otsu's thresholding [23], can address this problem. Thresholding is also called image segmentation, and the processed image is called a binarized image, as it is represented by two pixels only, viz., zero represents black and one represents white (see Fig. 3.5e).

5. **Rotation Correction:** An SEM image may be tilted, which could lead to cell segmentation failure. One solution is to apply Hough line detection. Once the space between rows of gates is detected (the line detected by Hough transformation [6]), we calculate the difference of the angle between the horizontal line and the detected line to correct image rotation.

3.4.5 Feature Extraction

The various well-developed 2D shape descriptors fall into three categories: contour, silhouette, and hybrid, the fusion of contour and silhouette. A contour-based descriptor can be used to make SEM feature representation and classification faster because contour descriptors reduce the number of pixels for calculation. 2D Fourier feature descriptors (FD) are among the most promising methods due to their simplicity and robustness to image rotation and noise. This feature descriptor is based on the Fourier transform. It is easy to implement and requires less computational effort than other contour-based descriptors such as the wavelet descriptor (WD) or the curvature scale space descriptor (CSSD). These FDs can easily describe the closed shape of p-doped and n-doped active regions connected

Fig. 3.6 Generation of
Fourier descriptors.
(**a**) Segmented SEM image of
logic cell. (**b**) Segmented
image. (**c**) Converting logic
cell into closed shape. (**d**)
Generating Fourier descriptor
for upper and lower closed
shape. (**e**) Combining upper
and lower descriptors for
whole logic cell [29]

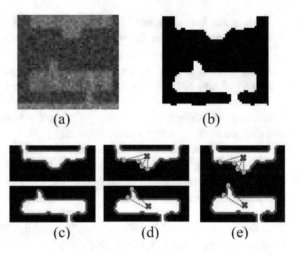

to the power and the ground rail of the circuit. FD for every logic gate (cell) can be
obtained through extracting contour coordinates from a closed 2D shape, calculating
shape signatures, and applying Fourier transform (see Fig. 3.6).

3.4.5.1 Elliptic Fourier Descriptor (EFD)

One of the most effective feature descriptors is the elliptic Fourier descriptor. It can
accurately describe a closed shape and is robust to rotation and scale. The idea of
EFD is based on the 2D Fourier descriptor and follows two steps: first, obtain the
contour coordinates pair from the cell silhouette and then apply the elliptic Fourier
expansion from the contour center based on four directional Fourier coefficients.

- Obtain contour coordinates: The initial contour coordinate (the 0th coordinate)
  is obtained by the pixel value difference on the cell edge; remaining contour
  coordinates are found using *chain code* [26].
- Elliptic Fourier expansion: The elliptic Fourier expansion is calculated by the
  center point of closed shape and four directional coefficients which are obtained
  with the following equations [10]:

$$a_n = \frac{T}{2n^2\pi^2} \sum_{p=1}^{k} \frac{d_{x_p}}{d_{t_p}} [\cos \frac{2n\pi t_p}{T} - \cos \frac{2n\pi t_{p-1}}{T}] \qquad (3.1)$$

$$b_n = \frac{T}{2n^2\pi^2} \sum_{p=1}^{k} \frac{d_{x_p}}{d_{t_p}} [\sin \frac{2n\pi t_p}{T} - \sin \frac{2n\pi t_{p-1}}{T}] \qquad (3.2)$$

$$c_n = \frac{T}{2n^2\pi^2} \sum_{p=1}^{k} \frac{d_{y_p}}{d_{t_p}} [\cos \frac{2n\pi t_p}{T} - \cos \frac{2n\pi t_{p-1}}{T}] \tag{3.3}$$

$$d_n = \frac{T}{2n^2\pi^2} \sum_{p=1}^{k} \frac{d_{y_p}}{d_{t_p}} [\cos \frac{2n\pi t_p}{T} - \cos \frac{2n\pi t_{p-1}}{T}] \tag{3.4}$$

where x and y are the pth coordinates, T is the perimeter of contour, d is the distance between adjacent coordinates pairs, and t is the distance between the pth coordinates and the 0th coordinates, and n is order of harmonics selected based on performance. The Fourier expansion is then given by

$$x_p = x_{center} + \sum_{n=1}^{N} [a_n \cos \frac{2n\pi t_p}{T} + b_n \sin \frac{2n\pi t_p}{T}] \tag{3.5}$$

$$y_p = y_{center} + \sum_{n=1}^{N} [c_n \cos \frac{2n\pi t_p}{T} + d_n \sin \frac{2n\pi t_p}{T}] \tag{3.6}$$

where x_{center} and y_{center} are the center coordinates of the contour area.

EFD can be made rotation-invariant and size-invariant through simple normalization. However, to detect Trojans that change cell size, we preserve the size information for each cell and only normalize EFD to be rotation-invariant. Since EFD describes a closed shape, logic cells need to be separated into their N and P regions, and then these regions must be grouped to produce a single descriptor for each cell. The new EFD for each cell c therefore becomes:

$$\text{EFD}_c = [\text{efd}_{N,c}, \text{efd}_{P,c}] \tag{3.7}$$

More coefficients require more calculation effort and will introduce noise into the same type of cell. Therefore, in this work, we take $n = 10$ harmonics (see Fig. 3.7) to approximate the shape of a cell. This results in 80-dimensional feature vectors of extracted gate footprints.

3.4.6 Image Classification

Image classification is the most critical step for physical inspection. This step involves grouping detected features into different subsets. One of the use cases for hardware assurance is the *golden gate* Trojan detection method. This approach compares trusted active-layer transistors (golden cells) with the rest of the transistors on a chip. Golden cells are fabricated on the same chip and can be quickly validated by a self electrical test for any malicious changes performed by an untrusted

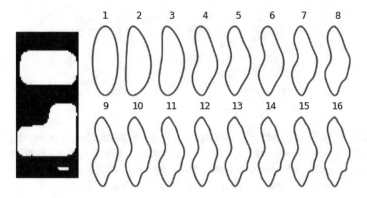

Fig. 3.7   Elliptic Fourier descriptor for a logic cell [27]

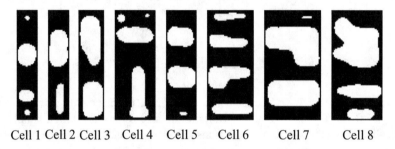

Fig. 3.8   Different logic cells identified from SEM image

foundry. If these cells (in the chip fabricated in an untrusted foundry) pass the test, they will act as on-chip golden reference cells for image analysis and classification to achieve very high accuracy in Trojan detection.

Classification can be performed by training a machine learning classifier, such as one-vs.-all support vector machines or a K-nearest neighbor classifier, to detect malicious changes. Reference images are typically compared to a golden image. A data set of different cells can be first prepared by clustering into different classes (see Fig. 3.8). These clustered cells can train a machine learning classifier and test against the rest of the imaged cell from the chip.

3.5   Terahertz-Wavelength Imaging: A New Modality for Physical Inspection

In the past two decades, significant advancements have been made in terahertz (THz)-wavelength photon generation, opening the door for novel surface and volumetric IC assurance methods. THz offers reasonable resolution images that can exploit unique material properties [3, 11].

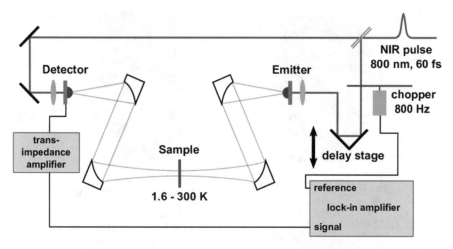

Fig. 3.9 Typical setup for sample analysis using a THz source. A beam splitter superimposes a reference copy of the source beam onto a version passing through the sample in question. For IC assurance, this would be the chip in question. (Retrieved from [21])

THz images are typically used to obtain sample depth information as well as material characterization analysis. Such information is collected by directing a beam of THz-wavelength photons at a target and examining the resultant transmitted and reflected energy spectra [2, 4]. An example lab setup is depicted below in Fig. 3.9. The method of THz generation can be adjusted to optimize for either frequency or time resolution [17].

3.5.1 Terahertz Applications in IC Assurance

Unique properties of THz imaging make it useful for several forms of IC assurance. As research further improves the cost, speed, and resolution of THz sources, THz will become increasingly valuable.

Trojan Detection Often, counterfeit ICs are coated, packaged, or modified such that they are indistinguishable from their genuine counterparts. As such, no form of surface-level visual imaging or microscopy will reveal the tampered component. However, THz imaging can obtain a subsurface rendering that exposes wiring within the chip package. This image can be compared against a known golden IC sample to verify the legitimacy of chips [5, 15]. A depiction of this comparison is given in Fig. 3.10. THz has similar capabilities to X-ray imaging but does not require sample setup or as costly imaging equipment [1].

Defect Detection For the reasons listed above, THz imaging is a useful tool for detecting defects within an IC sample. If the reflected and transmitted energy signatures for a known chip are characterized, readings for a chip that cannot

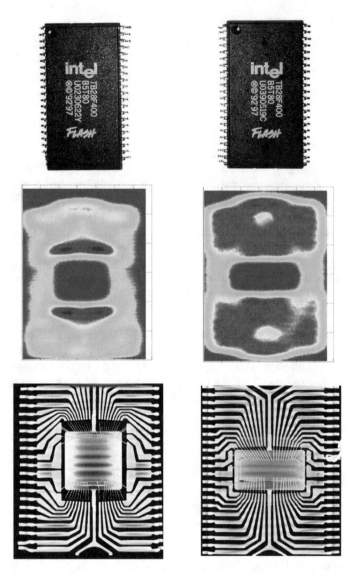

Fig. 3.10 Image comparison for authentic (left) and counterfeit (right) components. Top row: visual image. Middle row: THz image. Bottom row: X-ray image. Note that the stark contrast between counterfeit and authentic components is not evident in the visual image alone. (Adapted from [1])

pass quality control standards will be evident [22, 32]. Faults that affect electrical interconnects are particularly noticeable [16].

IC Fingerprinting THz imaging is highly sensitive to presence of water in a sample [14]. Figure 3.11 shows which illustrates the attenuation of THz waves

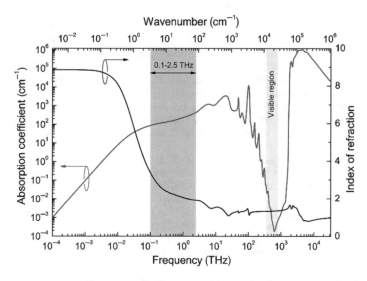

Fig. 3.11 Absorption coefficient of water molecules. Note the high impedance encountered in the THz regime in contrast to radio or visible frequencies. This can be exploited to perform sensitive measurements of packaging moisture content. Such a measure is useful for inferring sample age, potential defects, fingerprinting, and more. (Retrieved from [20])

due to water molecules compared to visual or radio wavelengths [20]. For ICs, this attenuation will vary depending on packaging conditions, the environment in which the chip was used, age, and other conditions. As a result, the attenuation profile for two samples of the same chip may be quite different. Such a factor can be exploited as a method of "fingerprinting" individual ICs and maintaining supply chain integrity [18, 19]. Fingerprinting can also extend beyond moisture and water content analysis. By collecting detailed profiles of material attenuation, unique information about sample packaging and internal variations can be quantified as well [12].

3.5.1.1   Future Applications

Novel methods for IC assurance continue to be developed with advancements in terahertz imaging. The dawn of THz optical coherence tomography (OCT) brings new potential to 3D THz assurance [8, 9]. Additionally, near-field imaging now allows for nanometer resolution sufficient for imaging current-generation transistor gates. Hence, this new technology may provide the means for assurance techniques such as p- or n-type doping detection at the transistor level [32].

References

1. Ahi, K., Anwar, M.: Advanced terahertz techniques for quality control and counterfeit detection. In: Terahertz Physics, Devices, and Systems X: Advanced Applications in Industry and Defense, vol. 9856, p. 98560G. International Society for Optics and Photonics (2016)
2. Ahi, K., Asadizanjani, N., Shahbazmohamadi, S., Tehranipoor, M., Anwar, M.: Terahertz characterization of electronic components and comparison of terahertz imaging with x-ray imaging techniques. In: Terahertz Physics, Devices, and Systems IX: Advanced Applications in Industry and Defense, vol. 9483, p. 94830K. International Society for Optics and Photonics (2015)
3. Ahi, K., Asadizanjani, N., Shahbazmohamadi, S., Tehranipoor, M., Anwar, M.: Thz techniques: a promising platform for authentication of electronic components. In: CHASE Conference on Secure/Trustworthy Systems and Supply Chain, Storrs (2015)
4. Ahi, K., Jessurun, N., Hosseini, M.P., Asadizanjani, N.: Survey of terahertz photonics and biophotonics. Opt. Eng. **59**(06), 1 (2020). https://doi.org/10.1117/1.OE.59.6.061629
5. Ahi, K., Shahbazmohamadi, S., Asadizanjani, N.: Quality control and authentication of packaged integrated circuits using enhanced-spatial-resolution terahertz time-domain spectroscopy and imaging. Opt. Lasers Eng. **104**, 274–284 (2018). https://doi.org/10.1016/j.optlaseng.2017.07.007
6. Ballard, D.H.: Generalizing the hough transform to detect arbitrary shapes. Pattern Recogn. **13**(2), 111–122 (1981)
7. Botero, U.J., Wilson, R., Lu, H., Rahman, M.T., Mallaiyan, M.A., Ganji, F., Asadizanjani, N., Tehranipoor, M.M., Woodard, D.L., Forte, D.: Hardware trust and assurance through reverse engineering: a survey and outlook from image analysis and machine learning perspectives. arXiv preprint: 2002.04210 (2020)
8. Dandolo, C.L.K., Brunel-Duverger, L., Giovannacci, D., Pillay, R., Lopez, M., Bai, X., Pagès-Camagna, S., Brodie-Linder, N., Menu, M., Detalle, V.: Terahertz time domain imaging and optical coherence tomography for the subsurface noninvasive inspection of a 21st dynasty Egyptian coffin. In: Liang, H., Groves, R., Targowski, P. (eds.) Optics for Arts, Architecture, and Archaeology VII, vol. 11058, pp. 174–181. International Society for Optics and Photonics, SPIE (2019). https://doi.org/10.1117/12.2527106
9. Dandolo, C.L.K., Lopez, M., Fukunaga, K., Ueno, Y., Pillay, R., Giovannacci, D., Du, Y.L., Bai, X., Menu, M., Detalle, V.: Toward a multimodal fusion of layered cultural object images: complementarity of optical coherence tomography and terahertz time-domain imaging in the heritage field. Appl. Opt. **58**(5), 1281–1290 (2019). https://doi.org/10.1364/AO.58.001281. http://ao.osa.org/abstract.cfm?URI=ao-58-5-1281
10. Diaz, G., Zuccarelli, A., Pelligra, I., Ghiani, A.: Elliptic fourier analysis of cell and nuclear shapes. Comput. Biomed. Res. **22**(5), 405–414 (1989)
11. Dorney, T.D., Baraniuk, R.G., Mittleman, D.M.: Material parameter estimation with terahertz time-domain spectroscopy. JOSA A **18**(7), 1562–1571 (2001). https://doi.org/10.1364/JOSAA.18.001562
12. Fischer, B.M., Wietzke, S., Reuter, M., Peters, O., Gente, R., Jansen, C., Vieweg, N., Koch, M.: Investigating material characteristics and morphology of polymers using terahertz technologies. IEEE Trans. Terahertz Sci. Technol. **3**(3), 259–268 (2013). https://doi.org/10.1109/TTHZ.2013.2255916
13. Gonzalez, R.C., Woods, R.E., Eddins, S.L.: Digital Image Processing Using MATLAB. Pearson Education India (2004)
14. Ho, L., Pepper, M., Taday, P.: Signatures and fingerprints. Nat. Photonics **2**(99), 541–543 (2008). https://doi.org/10.1038/nphoton.2008.174
15. Kawase, K.: For drug detection & large-scale integrated circuit inspection, p. 6
16. Kiwa, T., Tonouchi, M., Yamashita, M., Kawase, K.: Laser terahertz-emission microscope for inspecting electrical faults in integrated circuits. Opt. Lett. **28**(21), 2058–2060 (2003). https://doi.org/10.1364/OL.28.002058

17. Kong, D.Y., Wu, X.J., Wang, B., Gao, Y., Dai, J., Wang, L., Ruan, C.J., Miao, J.G.: High resolution continuous wave terahertz spectroscopy on solid-state samples with coherent detection. Opt. Lett. **26**(14), 17964 (2018). https://doi.org/10.1364/OE.26.017964
18. Luo, S., Wong, C.: Influence of temperature and humidity on adhesion of underfills for flip chip packaging. IEEE Trans. Compon. Packag. Technol. **28**(1), 88–94 (2005). https://doi.org/10.1109/TCAPT.2004.838872
19. Ma, X., Jansen, K.M.B., Ernst, L.J., van Driel, W.D., van der Sluis, O., Zhang, G.Q.: Characterization of moisture properties of polymers for ic packaging. Microelectron. Reliab. **47**(9), 1685–1689 (2007). https://doi.org/10.1016/j.microrel.2007.07.090
20. Moller, U., Cooke, D.G., Tanaka, K., Jepsen, P.U.: Terahertz reflection spectroscopy of debye relaxation in polar liquids [invited]. J. Opt. Soc. Am. B **26**(9), A113 (2009). https://doi.org/10.1364/JOSAB.26.00A113
21. Morris, C., Valdés Aguilar, R., Ghosh, A., Koohpayeh, S., Krizan, J., Cava, R., Tchernyshyov, O., McQueen, T., Armitage, N.: Hierarchy of bound states in the one-dimensional ferromagnetic ising chain conb2o6 investigated by high-resolution time-domain terahertz spectroscopy. Phys. Rev. Lett. **112**(13), 137403 (2014). https://doi.org/10.1103/PhysRevLett.112.137403
22. Naftaly, M., Vieweg, N., Deninger, A.: Industrial applications of terahertz sensing: state of play. Sensors **19**(1919), 4203 (2019). https://doi.org/10.3390/s19194203
23. Otsu, N.: A threshold selection method from gray-level histograms. IEEE Trans. Syst. Man Cybern. **9**(1), 62–66 (1979)
24. Rahman, M.T., Rahman, M.S., Wang, H., Tajik, S., Khalil, W., Farahmandi, F., Forte, D., Asadizanjani, N., Tehranipoor, M.: Defense-in-depth: a recipe for logic locking to prevail. Integration **72**, 39–57 (2020)
25. Rahman, M.T., Shi, Q., Tajik, S., Shen, H., Woodard, D.L., Tehranipoor, M., Asadizanjani, N.: Physical inspection & attacks: new frontier in hardware security. In: 2018 IEEE 3rd International Verification and Security Workshop (IVSW), pp. 93–102. IEEE (2018)
26. Rosenfeld, A.: Digital Picture Processing. Academic Press (1976)
27. Shi, Q., Vashistha, N., Lu, H., Shen, H., Tehranipoor, B., Woodard, D.L., Asadizanjani, N.: Golden gates: a new hybrid approach for rapid hardware trojan detection using testing and imaging. In: 2019 IEEE International Symposium on Hardware Oriented Security and Trust (HOST), pp. 61–71 (2019)
28. Tehranipoor, M., Koushanfar, F.: A survey of hardware trojan taxonomy and detection. IEEE Des. Test Comput. **27**(1), 10–25 (2010)
29. Vashistha, N., Lu, H., Shi, Q., Rahman, M.T., Shen, H., Woodard, D.L., Asadizanjani, N., Tehranipoor, M.: Trojan scanner: detecting hardware trojans with rapid sem imaging combined with image processing and machine learning. In: ISTFA 2018: Proceedings from the 44th International Symposium for Testing and Failure Analysis, p. 256. ASM International (2018)
30. Vashistha, N., Rahman, M.T., Shen, H., Woodard, D.L., Asadizanjani, N., Tehranipoor, M.: Detecting hardware trojans inserted by untrusted foundry using physical inspection and advanced image processing. J. Hardware Syst. Secur. **2**(4), 333–344 (2018)
31. Xiao, K., Forte, D., Jin, Y., Karri, R., Bhunia, S., Tehranipoor, M.: Hardware trojans: lessons learned after one decade of research. ACM Trans. Des. Autom. Electron. Syst. (TODAES) **22**(1), 6 (2016)
32. Yamashita, M., Otani, C., Kawase, K., Nikawa, K., Tonouchi, M.: Noncontact inspection technique for electrical failures in semiconductor devices using a laser terahertz emission microscope. Appl. Phys. Lett. **93**(4), 041117 (2008). https://doi.org/10.1063/1.2965810

Chapter 4
Physical Inspection of Printed Circuit Boards

4.1 Introduction

A printed circuit board (PCB) (usually with green color background) could be single- or multilayered and have a combination of active and passive components. It consists of connections between components using conductive pads and tracks [6]. PCB reverse engineering aims to analyze the circuit board, extract the circuit elements' connectivity, generate bill of material (BoM) file, and extract/evaluate functionality. Modern PCB reverse engineering is performed for a variety of reasons. For example, honest intentions encompass failure analysis, quality control, verification, fault identification, confirmation of the IP (intellectual property), tamper/counterfeit detection, and education. Alternatively, dishonest motivations include attack development, piracy, identification of vulnerabilities, cloning, and counterfeiting [16, 33, 34, 41]. Reverse engineering has always been considered a destructive, time-consuming, and expensive approach due to limited capability of imaging tools and expertise requirements. Recent advancement in image analysis incorporated with the improved and developed techniques relying on scanning electron microscope (SEM), X-ray CT, and optical imaging paved the path of developing more reliable, faster, non-destructive, and automated PCB reverse engineering methods [5, 11, 39].

Advancements in technology have led to PCBs becoming increasingly complex. Modern PCBs contain a huge variety of ICs, pins, components, and layers. They are designed by different design houses, according to requirements of the end users, and manufactured in various countries. Due to the shift of the PCB supply chain to a horizontal model to reduce the design and manufacturing cost, PCBs are becoming extremely vulnerable to diverse attacks and susceptible to root-of-trust violation, as shown in Fig. 4.1 [33]. For example, internal features and JTAG test pins can be accessible by unused sockets, probe pins, and USBs implemented in the PCB. Violation of the digital rights management (DRM) policy through JTAG in an Xbox is a real-life example of such susceptibilities of modern PCBs [36]. In

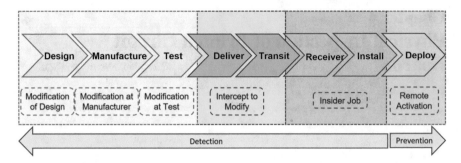

Fig. 4.1 PCB supply chain vulnerabilities

addition, due to extensive outsourcing of PCB manufacturing steps in the horizontal supply chain, hardware Trojan attack has become a widespread concern [9, 16]. One such attack was highlighted in a story named "The Big Hack" which was published on October 4, 2018 [35]. According to the Big Hack article published by Bloomberg, the security of the motherboards of Super Micro Computer Inc. (Supermicro) was compromised. Supermicro's motherboards, which functioned as the neurons of several data centers, were found to possess a small malicious chip that infected the motherboards with malware every time the server booted up. Though the investigation of the malicious chips began in 2015, it is believed that the attack on Supermicro's motherboards was initiated in 2014. Also, it is easier to reverse engineer a PCB to perform malicious modifications compared to ICs.

Reverse engineering is a process that could be performed at any level of electronic systems such as chip level, system level, and board level. Reverse engineering is also considered legal in many countries since the goals of reverse engineering include education, chip enhancement, reproduction, upgrading, and duplication. Reverse engineering (automated or manual) results in a netlist that can be used to reproduce the system. However, the benefits of reverse engineering can also be exploited by an unscrupulous entity. The adversary can exploit the acquired information to perform counterfeiting, cloning, Trojan insertion, and more as shown Fig. 4.2. Exploiting the knowledge gathered from PCB reverse engineering, adversaries could sell duplicates in large quantities without any actual development costs. Moreover, counterfeit components/PCBs could lead mission-critical systems to fail in-field, causing severe consequences, even endangering human life and national security. In addition, the root of trust of a system is violated once hardware Trojans or malicious modifications are inserted into the PCB design. Similar to the Big Hack incident [35], a single modification can lead to potential data leakage or significant performance degradation. Since PCBs are a fundamental component of any electronic system, it is important to verify the authenticity and integrity of the board itself.

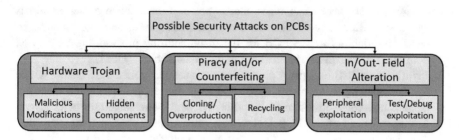

Fig. 4.2 Possible attacks on PCBs

4.2 PCB Life Cycle

The overview of the modern PCB life cycle is given below:

1. **PCB Designer:** The PCB designer performs the research, selection, simulation, board layout, verification, and validation. The completed design is passed to the design house with detailed design specifications. The design files are either generated by in-house engineers or third-party vendors.
2. **Design House:** As soon as the design files are obtained from the design engineer, board files are generated based on the netlist. Libraries for the components are created, and the data sheets are analyzed for all the components. Next, the placement of components is performed, and later net and interconnects are routed on the board. High priority is given to components with mechanical constraints as well as critical nets. At this stage, design rule checks are performed to eliminate errors. Finally, post-processing of the design is performed through which Gerber files, drill files, and assembly diagrams are created.
3. **Fabrication House:** In this stage, the boards are fabricated in the foundry. The assembly process begins once fabrication is complete. A detailed explanation of the assembly process follows in later Sect. 4.3. Later, the manufactured boards are sent for inspection. Common inspection techniques include automatic optical inspection and X-ray inspection. During the inspection, the PCB's integrity is automatically verified. After these tests, power-up tests such as in-circuit testing, JTAG testing, and functional testing are performed. The performance of the PCBs are evaluated based on different simulation environments. Faulty PCBs are repaired and verified to check the performance benchmarks. After this step, feedback is provided to design engineers.

4.3 PCB Assembly Process

1. **Establishment of the PCB design:** Softwares such as extended Gerber's are used to generate a blueprint of the PCB and encode all the important details.

Design aspect checks and error checks are performed once a Gerber file is generated. After these checks, the design files are passed onto the foundry. To check the viability of the PCB for the manufacturing process, the fabrication house performs a check called "Design for Manufacture" (DFM).

2. **Printing the PCB design:** A plotter printer is used to print a PCB design where for the inner layer, black ink denotes circuits and copper traces and clear ink denotes non-conductive materials and the inks are interchanged for outer layers. In this stage, all of the PCB layers with solder masks receive their own film. A registration hole is punched that is used as a reference point to align the films in the later stages.

3. **Copper for the internal layer:** In this first stage of the manufacturing process, copper pre-bonded to the lamination serves as a structure of the PCB, which is etched to reveal the blueprint. A photo-sensitive material named "resist" is used to cover the laminated panel. The goal of adding the resist is to receive a perfect match between the blueprint and the printer version. Laminations and the resists are now lined up, and they receive a blast of ultraviolet light. This process identifies the areas of copper, which will be used as pathways, and the rest of the photoresist is hardened. Later, PCBs are checked for errors. Later, the copper which was not covered by the photoresist is removed.

4. **Layer alignment:** In this stage, the optical punch is used to drive a pin through the holes to align the inner and outer layers of the PCB. Later, optical inspection is performed to check for errors. After the error checks, the PCB is considered ready for production. The layers are fused together and lamination process is begun. Pre-coated pieces of fiberglass with epoxy resin are used to make the outside of the PCB. The inner and the outer layers are put together using a metal clamp on a press table. A specialized pin is used to fit each layer. After these steps, the layers are pressed together. In the same stage, the pins are stacked down to ensure the layers are fixed correctly. Next, heat and pressure are applied to the stack of layers. Due to the heat, the epoxy resin melts and the pressure is used to fuse the layers of the PCB together.

5. **Drilling and PCB plating:** Drill spots are located using an X-ray machine. The extended Gerber files are used as referenced by the computer-guided drill to make the necessary holes. After that, all the layers of the PCB are fused together, followed by a chemical bath.

6. **Imaging outer layer and etching:** In this stage, the photoresist is applied only to the outer layer of the PCB for PCB imaging. The inner and outer layers are plated with tin. The tin serves as a protecting layer to protect the valued copper from being etched.

7. **Solder mask application:** Over cleaned PCB panels, ink epoxy is applied alongside a film of the solder mask. Later, ultraviolet light is blasted over the PCBs to identify the unwanted solder mask for removal.

8. **PCB silk screening, testing, and cutting:** PCB is plated with gold, HASL, or silver. In this silk screening process, the important information, such as manufacturing information, board IDs, and warnings, are printed. The boards are then sent for electrical testing. Later, PCBs are cut from their original panel.

4.4   Security Issues with PCBs

The modern horizontal business model offers the original component manufacturer (OCM) low manufacturing cost and reduced time to market through outsourcing PCB design and manufacturing process. As shown in the PCB life cycle (see Fig. 4.1), several untrusted entities are involved at different stages of PCB manufacturing. These untrusted entities include, but are not limited to, system integrators, individuals, groups, governments, fabrication facilities, and counterfeiting parties. Therefore, any interested dishonest party can leverage his/her access to the PCB manufacturing steps to launch stealthy attacks (e.g., backdoor or Trojan implementation in design) or gain unlawful advantages (e.g., IP piracy, counterfeit IC integration). A detailed description of these vulnerabilities is shown in Fig. 4.2, which exists in the supply chain. Thus, security of the PCBs is a difficult challenge. The IT industry is inexorably international, and anyone involved in the process could subvert the security of the end product. A few examples of possible PCB attacks are listed below.

4.4.1   Hardware Trojans: Kill Switch

A kill switch is used to cause a denial of service in the system. Destruction of entire databases, failure of networking systems, jamming weapons, or other critical infrastructures during emergencies or conflicts are possible implications of a kill switch.

4.4.2   Hardware Trojans: Infiltrate Communication Lines

Malicious circuitry implemented in a PCB can infiltrate a communication signal, cause unintentional alteration or corrupt, or path delay signals. The scenarios mentioned above can cause a product to fail in the field. Moreover, spies can use infiltrated communication lines to gain access to classified information.

4.4.3   Piracy/Counterfeiting: Cloning PCBs

An adversary with access to a pirated PCB design file can clone the PCB to make a profit. This drives business away from OCM that invested significant amounts of resources for research and development of the IP.

Unlike counterfeit electronics of the past, modern clones are very sophisticated [40]. These cloned PCBs could facilitate a nationwide attack. Clones, which have likely not undergone rigorous testing, are typically less reliable than the genuine products. Moreover, clones may host malicious software, firmware, or hardware, which could prove fatal in the field if the clones are used in security-sensitive, critical infrastructures.

4.5   Attack Models

Hardware Trojan attacks could be classified in the following categories. Category 1, the PCB design is trusted; and Category 2, the PCB design in untrusted. In scenario 1, the design house is considered trusted, the foundry/fabrication house is untrusted. The hardware Trojan inserted into the PCB gets activated under rare cases. Therefore, the PCB has higher probability of passing through functional and in-circuit testings. In scenario 2, both the PCB design house and the foundry/fabrication house are considered untrusted. In this scenario, the functional and parametric specifications of the board are trusted. In this scenario, an adversary has higher flexibility to alter or tamper PCBs or even insert a hardware Trojan in it. Again, these are activated in rare scenarios to avoid detection during testing. It is to be noted that in both the scenarios the motivation of the adversaries are either malfunction the PCB/system and/or information leakage. A few possible attacks are explained below.

4.5.1   In-Field Alteration

One of the most common and also the most dangerous attacks is in-field alteration. These alterations consist of altering components; replacing components; exploiting any available space on the PCB to add components; exploiting traces, vias, pins, and USBs; rerouting pads; soldering wires; and mounting ICs. One common example of PCB tampering is avoiding DRM protection, allowing the adversary to play illegally copied games and burnt, pirated, or unauthorized versions of the games and violate many other digital rights. The ultimate consequence of in-field alteration is revenue loss.

4.6   Current PCB Testing Approaches

4.6.1   In-Circuit Testing

In-circuit testing (ICT) involves the use of electrical probes to check the resistance, capacitance, or other electrical quantities at each node on a populated PCB.

This determines whether each component in the design file is placed correctly on the physical PCB. ICT can be performed using either a bed-of-nails test fixture or a flying probe setup. A bed-of-nails setup involves a test fixture with an array of small and spring-loaded pins [28]. Each pin makes contact with one node in the circuitry of the PCB. When pressed against the pins, contact is made with thousands of individual test points simultaneously. A flying probe test involves a pre-programmed automatically moving probe in a fixtureless setup.

4.6.1.1   Advantages

ICT can effectively detect defective solders, short circuits, open connections, and missing components.

4.6.1.2   Disadvantages

ICT is difficult to perform for densely populated, multilayered boards and can be expensive due to the costs associated with having to develop different test setups for different PCBs. Since ICT assumes that a correctly assembled board should work, it only tests the assembly and not the functionality of the PCB. Thus, ICT is ineffective for detecting connection faults, non-electrical defects, and hardware Trojans that do not affect the tested nodes.

4.6.2   Functional Testing

Functional testing is typically performed later in the PCB manufacturing cycle as a final check. A functional tester simulates the electrical inputs of the intended working environment, interfaces them to the PCB, and then measures the electrical outputs. Hence, functional testing evaluates the functionality of the product, independent of the assembly of the board. Functional testers typically consist of an interface to the PCB, a cabinet, cabling for the connection between all the instruments, a CPU, and monitors.

4.6.2.1   Advantages

Functional testing can identify functional defects, measure the PCB power consumption during operation, and uncover problems within the analog and digital circuitry.

4.6.2.2 Disadvantages

The functional testing setup is different for different boards; thus the exact hardware needed will vary depending on the PCB. Hence, to ensure high detection rates, it requires time, resources, software programming knowledge, and a deep understanding of the PCB and its working environment. A major drawback is that functional testing relies on connectors, which are prone to reliability issues due to wear and tear. Moreover, its effectiveness is influenced by the testing scope, simulated inputs, and how an "acceptable" output is defined. Hence, functional testing is ineffective for detecting hardware Trojans that do not alter the tested board functions.

4.6.3 JTAG Boundary Scan Testing

Boundary scan testing involves testing a PCB's integrated circuits and wire lines through a JTAG test access port. Though the JTAG boundary scan does not test all nodes, it provides the equivalent of ICT without the use of fixtures.

4.6.3.1 Advantages

JTAG allows the reuse of test patterns from system level to board level and from board level to chip level. Since boundary scan testing is fixtureless and does not test all nodes, it is less expensive and faster than ICT.

4.6.3.2 Disadvantages

Including a JTAG test access port on a PCB increases the cost and area overhead. Since the JTAG boundary scan only tests chips structurally and PCB traces, it is ineffective for detecting hardware Trojans that do not affect the tested nodes.

4.6.4 Bare Board Testing

Bare board testing involves checking if circuit connections appear as noted on a bare circuit board. This process requires a netlist and a multimeter to perform a continuity test. These determine short circuits and open circuits by "charging" a net and then probing each net to measure the induced capacity.

4.6.4.1   Advantages

This type of testing is inexpensive and easy to perform in comparison to the other testing techniques listed above.

4.6.4.2   Disadvantages

Since a bare PCB board is evaluated, neither the function nor the assembly of the final product is tested. Therefore, bare board testing cannot be used to detect faults or malicious implants that occur during assembly. Further, unpopulated versions of the boards and netlists may not be available.

Testing technique	PCB stage	Human factors	Hardware Trojans detected	Cost
In-circuit testing	Populated boards	Bed-of-nails development/flying probe test programming; board transport for test setup	Trojans that alter tested nodes	≈$20k+
Functional testing		Software programming; board transport for test setup	Trojans that alter tested board functions	≈$50k+
JTAG testing		Higher design requirements; software programming; board transport for test setup	Trojans that alter tested nodes	≈$119+
Bare board testing	Unpopulated boards	Board transport for test setup	Cannot detect Trojans	≈$2k+
Visual inspection	Populated/unpopulated boards	Software programming; potentially no board transport for test setup automation possible	With multiple imaging modalities, detect all types of Trojans	≈$12k+

4.7   Trust and Hardware Assurance Through Reverse Engineering

4.7.1   Preliminaries

The fabrication of individual microelectronic components and final assembled devices takes place in several geographical locations from countless vendors across the globe. The electronic supply chain is defined by its horizontally integrated business model, resulting in the compartmentalizing of each aspect of the process from the design through fabrication to final test and sale of the device. This segmenting of the supply chain results in business specialization, which decreases total fabrication costs and increases the quality of services provided for a consumer.

Due to the untrusted nature of the supply chain, hardware security experts are investigating novel verification tools and the tracing of components with techniques such as block-chain for hardware assurance [45]. The current methods of complete and effective hardware assurance require a combination of non-destructive analysis on batches of samples and destructive analysis on individual samples to develop a high level of confidence in the functionality. While destructive testing offers high-quality internal analysis comparative to non-destructive testing, the sample preparation methods required create inherent damages and are tedious to perform. Sample preparation often destroys the device, and thus, it cannot be sold or used afterward; for this reason, non-destructive analysis with the internal spatial resolution, and thus confidence, of destructive methods is needed to protect end-user electronics. Electronic waste recycling is often used to monetize the elemental constituents of a PCB or IC [13]. The industry growth of electronics recycling also poses an increased vulnerability and economic incentive to recycle, counterfeit, and monetize entire PCBs and ICs which no longer have trusted functionality, but resemble OEM devices.

4.7.1.1   Application

Each of the various entities in the supply chain has an economic motivation to maximize their own profitability and revenues. This is due to their limited expansion capabilities in the horizontally integrated supply chain of electronics production. For this reason, the lowest-cost method of fabrication from the quality of supplies to the skill of labor will be employed to minimize costs. In some cases, the reduction of costs for one entity in the supply chain can result in increased operational costs for another. For example, a final testing entity facing a global supply chain must further verify the final functionality and safety of the shipped devices due to an increased likelihood of faulty or malicious operation. This might require extensive testing through functional, circuit, and volumetric methods to verify that the device is operating to specification and is free from modifications. The current methods of verification are unable to ensure a completely assembled device non-destructively and require destructive micro-sectioning methods to guarantee assurance.

Contrasting Challenges of Reverse Engineering: PCB vs. IC

PCBs are the electrical carrier for many electronic components from integrated circuits (ICs) themselves to optical sensors and more simple components like resistors. Due to the mass producibility, mechanical rigidity, and simplicity of electrical connectivity offered by PCBs, they are found anywhere from consumer electronics to industrial equipment. The components are embedded electrically and physically into a "grid" of copper traces, laterally by means of vias, and vertically among the PCB. This grid facilitates the interconnection of components into a functional device but also exposes the encrypted or unencrypted signals of the interconnected sophisticated components. Regardless of the type of microelectronic, it is possible for adversaries to reverse engineer any electrical circuitry with the appropriate metrology tools and time, at either the IC or PCB level.

Due to the larger-sized features on a PCB, it is an order of magnitude easier in cost, skill, and time required to attack through modification or duplication of a PCB compared to an IC. Integrated circuit vendors are constantly updating their design to improve specifications, and the market is very competitive. This along with the growing need to pack more components together to create high-volume sophisticated end-user devices such as smartphones has led to making the ICs as small as possible. The minimum feature of an IC can be on the order of nanometers for processors and memory, while PCBs' internal traces, spacing, and board thickness are all on the order of micrometers or microns. For this reason, the cost of metrology and characterization equipment to verify and/or attack an IC is vastly more expensive than a PCB and makes PCB especially vulnerable as they carry and transmit signals for individual components and the entire functional device itself.

Hence, it can raise serious issues concerning intellectual property (IP) infringement, the (in)effectiveness of security-related measures, and even new opportunities for injecting hardware Trojans. Ironically, reverse engineering can enable IP owners to verify and validate the design [9].

Untrusted Use of Design-Related Intellectual Property

Outside of design and testing, there is little variation and complexity in the assembly of a PCB. PCB fabrication has been outsourced to the Asia-Pacific region as operational costs are reduced due to labor, currency, and legal disparities compared to Europe and North America. This opens the door for potential adversaries such as manufacturers to implement lower-quality components, sell unauthorized copies, or more severely implement a hardware Trojan into a PCB during fabrication. The design knowledge required for fabrication enables manufacturers to more successfully modify a device without subsequent detection. This is when compared to the limited knowledge available to a design-for-test vendor or end users attempting in-field modification.

IP has become more difficult to enforce across borders, and the global nature of the supply chain has created gray areas of operations for malicious actors without the fear of prosecution. It is difficult to enforce the misuse of IP once it is provided to foreign entities as the Intellectual Property Rights Index for each country. The WTO has developed this Index in order to characterize the compliance with its regulations, and offshore entities have lower compliance levels. The inherent incentive to maximize revenues, whether legitimate or not, can result in the malicious reverse engineering of PCBs during production or once the PCB is available to the public. The driver for this type of malicious activity is to profitably derive privy knowledge of design for duplication or recreation for sale on the free market or in the potential case of a nation-state, for use as a zero-day hardware exploit. In either case, this can be done through obtaining of the bill of materials and then the circuit architecture otherwise known as the netlist. This process is very time-intensive and costly if it is required to start from scratch which is often the case of in-field modifications. This time to completion can be simplified by garnered knowledge leading to a faster return on its reverse engineering investment, and this is the advantage and threat of a manufacturer with supplied design knowledge.

The implications of the supply chain vulnerabilities on consumers and governments can lead to massive loss of revenue or more critically loss of life. The possibility of compromising of a corporate business model or a physically protected military asset through a PCB modification has increased in scope with the global expansion of the electronic supply chain. Similar to how corporate entities have an incentive to maximize revenues, government entities conducting statecraft have an incentive to compromise opposing states, whether friendly or hostile. Due to the widespread daily usage of electronics by consumers, corporations, and government entities, there is a large need to verify these devices for their electrical and mechanical functionality. The verification of an electronics' functionality is essential for the failure analysis of production errors or detection of malicious modifications, which both can cause issues during a device's life cycle operation.

4.7.2   Destructive vs. Non-destructive Methods for PCB Verification

The complicated interconnection of various vendor's components and the course of scaling up a prototype design into mass production will often lead to process issues for PCB manufacturers. These process issues can be due to copper traces, faults in a vendor's surface-mounted components, solder issues, etc. They can all result in failure at the device level. This type of failure is quite common and is mitigated through in-line electrical, physical, and optical inspection to prevent defects and maximize yield. These methods are employed on all production samples and thus are exclusively non-destructive and must perform their analysis in seconds. The additive nature of the layer-by-layer PCB production process can result in compounding of errors if a process error remains undetected for a period of time and continues to be added upon. In order to mitigate compounding of errors and maximize yield, the assembly line of a PCB will often include metrology steps to verify the process and validate if a device is ready for continued production. These in-line tools are only able to verify a component to a certain level, while it requires extensive destructive micro-sectioning of PCB devices to fully characterize their internal dimensions and components. The destructive methods have a higher resolution but require more time through analysis and sample preparation and inherently destroy the sample under test. To effectively analyze a PCB destructively, it will require many sacrificial samples to develop the process to derive all necessary knowledge.

4.7.2.1   In-Line Tools

Manufacturers employ both in-line and off-line methods to ramp up their production while maximizing yield and ensuring a level of functionality of their device to meet the customer spec required. The customer which might be the original designer and entity that completes the final sale of the device could be, for example, a tech giant like Apple or a transportation magnate like Ford. This customer may also employ similar in-line characterization on each product or randomized statistical selection from batches for off-line techniques to further verify the functionality and thus safety of their outsourced device. It is important for all PCB designers to ensure their device will operate effectively over the course of its lifetime and does not contain any process errors or malicious insertions that will affect the PCB's functionality. This can ensure the safety of their revenue stream and also their customer base.

As the sophistication of PCBs increased over time, there has been a shift from the manual visual inspections and personnel-intensive in-circuit testing toward automated optical methods and electrical test fixtures. This has resulted in a large increase in yield optimization for PCB manufacturers. There are many in-line tests performed in between steps of production, such as in-line functional testing and automated optical inspection. These verifications ensure a level of confidence for each of the samples' functionality and operation, as the length of a PCB life cycle can be limited by poorly processed electrical connections.

Electrical tests are limited by noise due to high levels of process variation between devices, while optical tests are hindered by densely packed components, multilayer boards, and variations in lighting conditions. The advent of surface-mount technology or SMT simplified the PCB production process and increased the usage of pick and place tools, which reduced defects due to handling errors. SMT enabled an increase in circuit density compared to through-hole technologies [27].

Although SMT enabled increases in final yield, there was a need to see behind the SMT component's packaging. This resulted in the need for an in-line non-destructive volumetric characterization tool to image these buried connections. One widely known form of volumetric characterization available in industry is automated X-ray imaging or AXI. AXI can be categorized into three areas based on its capabilities 2D, 2.5D, and 3D imaging, which is covered in off-line volumetric. These are traditional, tilted view, planar, and computed tomography, respectively [8].

Two-Dimensional (2D) X-Ray Imaging

Two-dimensional (2D) X-ray imaging can be considered a top-down view of the assembly line using a line scan or planar detector. This is a low-resolution method of analyzing defects such as disconnections by using 2D X-ray imaging, but can characterize a large number of process defects. For its low resolution, it is also low cost enabling in-line monitoring of various connections.

Tomosynthesis (2.5D) X-Ray Imaging

Titled view X-ray imaging combines the geometric benefits of 3D imaging with the cost-effectiveness and simplicity of traditional 2D imaging. The result is an application-specific trajectory that enables viewing "around" structures to see behind them. This is accomplished by tilting the source and detector relative to the planar PCB sample, and this angle assists in the viewing of deeply embedded structures hidden by dense material such as ball grid array connections. While 2D and 2.5D have high throughput, their low resolution and a small imaging area result in very limited applications. 2D and 2.5D are capable of non-destructively imaging embedded electrical connections, but their limited field of view and resolution are not enough to characterize an entire PCB volume. 2D and 2.5D are sufficient for PCB manufacturers to mitigate process defects and ensure adequate device operation for their customers, but not enough for complete non-destructive reverse engineering.

4.7.2.2 Off-Line Tools

While in-line tools are non-destructive, fast, and automated, off-line metrology tools and methods require skilled labor, are time-consuming, and are often destructive. If a complicated process error is not able to be mitigated through the in-line or on-line metrology tools, sometimes there is a need for failure analysis via an off-line inspection. Off-line metrology has the advantage of higher resolution that can highlight defects not visible during production and enhance a specific region of interest. However, these methods require extensive sample preparation, which adds significant time and results in the sacrificial loss of a production sample. The trade-off is a higher-resolution destructive method that might not be accessible during production to investigate an issue with electrical, volumetric, and/or mechanical tools. These characterization techniques can be combined with in-line production data to isolate a process issue and quickly adjust manufacturing parameters appropriately.

Volumetric (3D) X-Ray Imaging

There is a much larger cost associated with 3D X-ray imaging systems, and they often do not have the throughput on the order of seconds required for in-line monitoring. However, planar tomography using either laminography or tomosynthesis with novel geometries or limited step sizes is possible to quickly reconstruct representative 3D usable information in a high-throughput volume [12]. These methods require accurate sample stage control and high energy sources compared to 2D methods, but their simulated 3D resolution cannot compare to 3D computed tomography or CT. This is because CT captures 2D images from anywhere from 500 to 3000 on average projections and combines them into a volumetric representation. However, this complete imaging is very time-consuming to collect the data required for characterizing an entire volume with millimeter resolution.

This imaging method is similar to AXI. However, to obtain the millimeter volumetric resolution necessary for automated segmentation of internal PCB features, the current throughput for micro-CT is on the order of hours not seconds. Micro-CT is an off-line process used on test samples and is not currently capable of automatically verifying entire PCBs without skilled user intervention. This illustrates the growing need for an automated high-throughput, in-line, 3D X-ray micro-computed tomography (X-ray u-CT) to expand the non-destructive inspection of PCBs. There is growing work in reconstruction methods that can implement prior information and leverage the use of machine learning to enable increased resolution and throughput.

4.7.3 Reverse Engineering

Reverse engineering or RE of a PCB requires a multi-level approach: the classification of surface-mounted components, extracting the internal electrical structure, and fine-tuning the functionality such as the firmware. Each of these levels is moderately difficult but can be optimized with automation and augmented by access to sophisticated metrology tools and experienced labor.

RE will often require physical access to a targeted device aside from possible cyberespionage of the design schematics or firmware. Physical access can be during production for a manufacturer or in the field for an end user. This physical access enables various testing methods unavailable if a target device is only accessed remotely. Also, more than one target sample is often necessary in order to acquire various facets of physical and electrical design. It is not easy to destructively reverse engineer a PCB with only one sample as the multi-level approach requires many sacrificial samples. This creates a barrier to entry for RE because multiple target samples must be first prepared mechanically or chemically before they can be effectively characterized via optical or electrical methods. Each sacrificial sample provides additional knowledge specific to the PCB, which is used toward a final development of the multi-level RE approach for that individual PCB.

Comparatively, non-destructive test methods can evaluate the sample without causing damage enabling repetitive or thorough testing without requiring multiple sample devices. This results in minimizing the required labor for sample preparation and the associated characterization noise errors introduced via destructively polishing or etching. There is a significant concern in the non-destructive threat of RE as it can be performed while requiring only one sample and does not destroy the sample enabling repetitive tests or the return of the device to regular operation.

4.7.3.1   Potential Adversaries for PCB

Due to the widespread industry usage of in-line characterization and off-line failure analysis tools, they have become relatively cost-effective for their advanced automated and resolution capabilities. The skilled labor required to perform failure analysis to ensure proper device operation also enables the ability to derive privy knowledge from PCBs cost-effectively. This creates the increased vulnerability for PCBs to be simply reverse engineered for profit or malicious purposes. It is now easy to reverse engineer one- and two-sided PCBs with automated optical methods, and soon the combination of volumetric analysis will enable RE of multilayered boards with ease [6]. Thus, there is a drive to conceal hardware designs without compromising the capability to consistently verify individual components and overall PCB functionality, all within a cost-effective framework.

Manufacturer

Manufacturers are the highest level of threat to trust assurance as they have access to and require proprietary design and firmware knowledge to produce and test a device. It is difficult to ensure that a manufacturer will not duplicate a design for resale, and thus it is important to design a PCB to prevent the immediate derivation of its operation through knowledge of the schematic. A manufacturer is capable of implementing a hardware Trojan that might be undetectable until put into operation and can as simple as an addition of a line of circuitry as seen in Fig. 4.3.

In addition to access, the manufacturer is required to have the PCB design criteria in order to appropriately fabricate the PCB device. This step of the development cycle of a PCB is highly vulnerable to Trojan insertion, due to the design knowledge received by the manufacturer, which enables the implementation of a Trojan not detectable by non-destructive techniques such as functional testing or side-channel analysis during post-production. Manufactures have access to in-line tools and off-line metrology tools in order to appropriately characterize device faults during their production process. These off-line tools enable failure analysis, but also the high-level analysis required to maliciously implement an undetectable hardware Trojan into a PCB device.

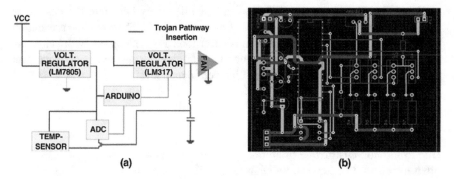

Fig. 4.3 (**a**) An Arduino fan controller circuit depicting an example of a Trojan insertion by a manufacturer through a trace (red) insertion that would cause failure of the temperature sensor, and subsequently the fan would fail to function based on temperature. (**b**) A two-layer PCB layout of the original circuit. (Retrieved from [14])

End Users and Competitors

Consumers or competitors can gain access to a target component through the free market and extract the critical targeted electronics for RE. In certain applications, such as military weapons, satellites, and internal company hardware, the device can remain protected from an unsophisticated adversary obtaining physical access throughout its life cycle. These entities often lack the resources and technical understanding necessary to reverse engineer a design quickly. This is because the design criteria is unknown to the adversary. Therefore, she needs access to extensive failure analysis tools and netlist extraction methods. However, with access to failure analysis tools and capability of reverse engineering, a competitor can be the most dangerous adversary because this entity can learn the IP secrets from any product available in the open market.

4.7.3.2   Assurance

In order to assure a level of confidence in the function of an electrical device, each one is often put through a functional test to simulate the conditions the device will operate under and check for potential failures. This test can be performed by the fabricator, a design-for-test (DFT) vendor, or even the end user and is meant to trigger any defective component from a low-quality supplier, indicate poor fabrication specs, or detect potential absent or implanted components. Functional tests and other inspection methods provide verification of the components, circuitry, and overall operation which gives a level of confidence for the inspected device to function as intended over its life cycle. This confidence level is related to the resolution of the various inspection methods implemented to verify design and functionality. These test methods, however, are unable to detect or mitigate a malicious insertion or removal if the attack is effectively built into the base level design of the circuit circumventing a detectable test metric. The complete prevention of all process errors and all malicious insertions is challenging with only one modality of testing, but the combination of modalities can ensure an adequate level of cross-over in resolution enabling a high confidence level.

4.7.3.3 Micro-sectioning

Due to the non-destructive limitations of functional and volumetric analysis, it is often necessary to destructively analyze a PCB. This method of complete internal volume analysis via sectioning allows for absolute confirmation of design to the acquired netlist from RE. However, this is a time-consuming process that requires a sample, and often many more, to be destroyed. This prevents the further use of this device and does not assure the safety of other devices not destructively analyzed. Although this method is not cost-effective, it is currently the only method of assuring hardware is free entirely of modifications or insertions. There are errors that can be introduced via sectioning destructively that are difficult to avoid and leads more noise and add complexity to the RE process. As seen in Fig. 4.4, the internal structure of a PCB is complex, and there are many connections and components that must be fabricated to exact specifications for functional operation.

4.7.4   Automated Techniques for Hardware Assurance

The current methods for PCB reverse engineering or RE require sophisticated metrology equipment and the employment of highly skilled labor. Expensive metrology tools from electrical oscilloscopes, optical microscopes, and nanoprobes and a skilled labor force with hands-on experience using these tools for failure

Fig. 4.4 (**a**) Surface view and (**b**) side view of a multilayer PCB implemented with inter-layer connectivity

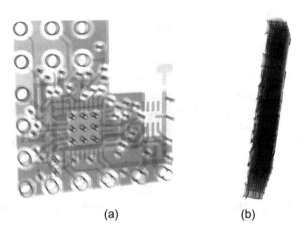

(a)   (b)

analysis are some of the cost limitations for reverse engineering. The implementation of automated techniques that can adaptively perform this manual preparation, metrology, and analysis can reduce the cost requirements for specialized tools and payroll.

Automated techniques are repeatable, unlike human labor, and can eliminate human errors that manifest into the design during sample preparation or circuit interpolation. The goal of reverse engineering is to develop the netlist and bill of materials of a device under inspection. The current methods of automated reverse engineering employ a combination of manual and automated techniques that analyze a PCB surface to derive a bill of materials and the circuitry. These surface components and the exposed circuitry can be optically imaged, filtered, segmented to separate traces and vias, and plated through holes, while components are analyzed via Internet search using their visible inscription. For more sophisticated designs, a volumetric analysis is needed for boards with internal layers, which can be performed destructively layer by layer or using X-ray or acoustic techniques.

Automating the reverse engineering of a printed circuit board without any given design information is very difficult, as all digital information must be derived from physical samples. This often requires multiple samples and is performed destructively in series to find the bounds and limitations necessary to image and derive all information for re-creating a design. This scenario is the case of an attacker or malicious actor, and this process is most difficult to automate when compared to the case of a manufacturer or designer. In either of these two cases, some or all of the design information is known, which can be used to feed test points to automated flying probe methods. These are capable of automatically testing connections and springing a netlist of electrical outputs that can be compared to the expected values. Designers and manufacturers can more easily implement automated techniques because of their access to the design information.

Testing methods for hardware assurance are often employed to confirm the design of sophisticated PCB designs, which requires the use of multilayer boards. Due to this, a volumetric analysis is often necessary, and current X-ray imaging resolution is often impaired by noise generated by the PCB. For this reason, destructive techniques such as micro-sectioning are used to confirm that a PCB design has been fabricated to specification. This testing can reveal process errors or malicious additions while comparing electrical functionality to design. Compared to an attacker destructively reverse engineering a design, a designer or fabricator can more effectively analyze their device under test for failure analysis by understanding the complete design.

4.7.4.1   Automated RE Methods

Due to the difficulty and time-intensive process of verifying all PCBs through either non-destructive or destructive methods, there is a need to develop a high-throughput guaranteed method of validating PCBs at various stages of the supply chain and during life cycle operation. The noise levels associated with process variations inhibiting electrical test effectiveness and the limited resolution of two-dimensional X-ray imaging are both incapable of the accuracy needed for verification. Also, imaging, reconstruction, and further user inputs needed to implement industrial X-ray tomography as a verification tool are not up to the speed of the high volume capacity needed to verify batches of PCBs.

These limitations require a need for high-throughput inspection that is capable of accurately detecting surface-level components along with internal volumetric trace, via, and embedded devices. The sophistication of devices calls for multiple metrology methods to be combined in order to derive the surface features using optical methods and combining this with internal volume information to verify the entire device. The current state of the art for automated reverse engineering requires the removal of surface components before volumetric imaging with lab source X-ray imaging. The noise resulting from surface components currently limits complete non-destructive analysis and verification of a PCB to its design file.

4.8 Taxonomy

There is a vast amount of process errors and hardware Trojans that can occur when a PCB is fabricated or during final usage. The process errors can range from trace and via sizes not meeting the specification for electrical current to voids in solder connections between components. Meanwhile, the sophistication of the hardware Trojans can range from designer or fabricator implementations to more simple in-field modification by end users as seen with the Bloomberg Big Hack and the malicious modifying of PCBs via test structures, respectively [42]. A comprehensive taxonomy based on areas of focus for PCB hardware assurance seen in Fig. 4.5 has been compiled along with the tools used to mitigate process errors and hardware Trojans [17].

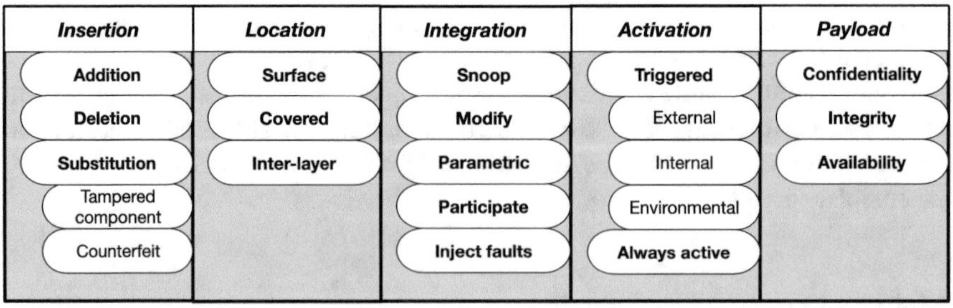

Insertion	Location	Integration	Activation	Payload
Addition	Surface	Snoop	Triggered	Confidentiality
Deletion	Covered	Modify	External	Integrity
Substitution	Inter-layer	Parametric	Internal	Availability
Tampered component		Participate	Environmental	
Counterfeit		Inject faults	Always active	

Fig. 4.5 Taxonomy of system-level Trojans

4.9   Image Processing, Computer Vision, and Machine Learning

Classifying Trojan taxonomies and PCB supply chain vulnerabilities is only the first step in end-to-end hardware assurance. The second portion involves acting on such knowledge to locate defects or tampered components, which finally distinguishes boards at the sample level. To perform a separation of verified boards from tampered or defective units, a combination of image processing, computer vision, and machine learning must be employed. Taken into account along with input from a human subject matter expert (SME), this procedure can be established. Typical considerations and imaging modalities encountered in non-destructive hardware assurance are discussed in the following sections.

4.9.1   Computer Vision and Image Processing

Image processing and computer vision (IP/CV) are intimately related. While image processing refers to lower-level operations, computer vision entails the higher-level knowledge gained about an image. Generally, image processing operations consume images as inputs and output image-like data. Computer vision operations use images as inputs and produce knowledge or system/object information as an output [15].

Consider an example where product manufacturers wish to find scratches on a specific type of IC. Image processing algorithms would "clean" the input image, making such defects stand out more clearly. Computer vision processes would then yield information, for example, scratch length, orientation, and other information SMEs can utilize to mark a chip as defective [1].

4.9.2   Comparison to Machine Learning

Machine learning (ML) works at an even higher level than computer vision. Taking in either images or knowledge as input in the form of *features*, it then produces an output expressing the relationship between these inputs. This relationship can subsequently be exploited to better characterize defects or Trojans for a given set of samples. ML techniques are not independent from IP/CV algorithms. Rather, the latter are used in *preprocessing* stages to gather the features examined by an ML method [38]. The relationship between each of these concepts is depicted in Fig. 4.6 and Table 4.1. Moving from specific (ML) to broad (SME verification) across the table, it is clear the former is heavily relied upon for providing automated hardware assurance.

ML methods can be split into two groups depending on the information they require to learn input relationships. These categories are *unsupervised* and *super-*

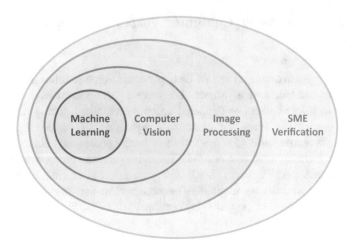

Fig. 4.6 Hierarchical relationship between machine learning, computer vision, image processing, and general SME verification. In this case, machine learning represents the highest level of complexity, while SME verification has the broadest applicability

Table 4.1 Important characteristics about the relationship between machine learning, computer vision, image processing, and general SME verification. The colors and titles of each column correspond to entries in the Venn diagram of Fig. 4.6

Stage Name	Machine Learning	Computer Vision	Image Processing	SME Verification
Used For	Sample Classification	Feature Extraction	Preprocessing	Multipurpose
Characteristics	• Can be fully automated • SME only needed for periodic random sampling • Widest range of applicability	• Start of automated assurance • Output no longer resembles an image • Only minor SME input required	• Clean noise in input image • Enhance appearance of defects • Still requires SME input	• Burden of assurance is completely on the human • Only useful for low-throughput supply chains • Highly prone to error

vised learning. Some techniques fit into the intersection between these, forming an additional group of *semi-supervised* methods.

4.9.2.1 Unsupervised ML Methods

Recall the objective of automatically detecting scratch defects on manufactured ICs. IP methods would first be used to reduce noise from these images, such as blurring, salt-and-pepper effects, white noise, and more. Next, CV techniques would extract features characteristic of scratch defects. These might include the number of sharp edges in the image, their orientation, brightness dispersion across the IC surface, and many other parameters. In an unsupervised network, these values are fed into an ML method which looks for natural groups or *clusters* in the data. Ideally, each cluster will indicate a *class* of IC, either defective (scratched) or normal. However, the unsupervised ML method is not capable of outputting this assessment directly; it is up to the researcher to provide a class label to each cluster.

4.9.2.2   Supervised ML Methods

Supervised ML techniques generally follow the same procedure and requirements as unsupervised learning for data collection. However, these methods *also* need an associated output value for each set of inputs. In the scenario outlined above, this would mean collecting the class of IC (scratched or normal) as well as all the previous features. This *labeled data* (inputs with known associated output values) allows for a much wider variety of ML techniques compared to unsupervised learning. Rather than only determining data clusters, supervised techniques are capable of automated regression and classification. That is, no further information is needed from researchers to output the class value directly.

4.9.2.3   Semi-supervised ML Methods

As the name suggests, this group consists of a pairing of techniques from the previous two sections. In such cases, only a small amount of labeled ground truth data is fed into the machine learning algorithm, which proceeds to classify large groups of unknown data.

4.9.3   Uses in PCB Assurance

CV/IP techniques can be used to provide some level of hardware assurance without the assistance of further processing methods. However, they are typically used in conjunction with machine learning or further statistical analysis to solve more complex assurance scenarios.

Cheap, Fast Quality Assurance [7, 30, 32]   Image processing and computer vision algorithms excel at rapid and low-cost assessment on a product line. Moreover, these techniques scale quite well with increasing computing power. As such, they are the baseline for verification and quality assurance metrics.

However, this speed comes at the cost of limited cross-applicability. It is difficult, if not impossible, to transfer computer vision verification methods from one product line to another without re-evaluating the characteristics of the manufactured items. This is because such methods are highly sensitive to the images themselves. Consequently, manufacturers with low-volume, high-variety yields must consider alternative options.

Tamper Detection [20, 22]   Beyond quality assurance, IP/CV methods offer the ability to detect Trojans and other malicious alterations. However, these options are limited due to the wide variety and constantly changing nature of such attacks. Hence, tamper detection is limited to product lines with few on-board components, high-precision manufacturing, and minimal changes across product generations.

Verification Enhancements [18, 43] As previously explained, ML techniques are generally not used as standalone processes; rather, they are employed as methods for fault detection and localization after features are collected by CV/IP algorithms [19]. However, unlike CV/IP bases assurance, ML methods allow for far more advanced fault and tamper detection[26]. They often work without the use of a golden sample and are far more robust to outlier boards or unique imaging conditions— provided a large enough training database is available. Moreover, this additional layer of complexity allows ML-based approaches to function in reverse engineering applications.

Unfortunately, an extraordinarily large database is required for consistent accuracy in large-scale hardware assurance. Due to the widely diverse nature of PCB samples, any small database will not contain enough information about any one board design to properly characterize its unique features. However, for localized product lines employing rigorous imaging conditions, this condition is not necessary.

4.9.4 Verification Process Flow

Without the addition of further processing techniques, CV/IP methods require a golden sample to provide hardware assurance. Due to the complex process of PCB manufacturing and wide variety of PCB types, no general metric can be applied to verify a PCB without a reference board. Most CV/IP techniques for hardware assurance follow the same procedure, whose stages are outlined below.

4.9.4.1 Preprocessing

On receiving an input, fault/hardware Trojan detection algorithms begin by preprocessing the image. This involves running a host of operations to remove as much noise as possible and prepare the image for meaningful difference imaging. Without this stage, the difference imaging stage will reveal too many false positives to output meaningful information.

The type of noise present in an image varies widely depending on multiple parameters, such as image resolution, modality, equipment quality and age, technical skill of the operator, and more. As such, an extensive set of tools is available to address each of the potential noise sources in the input image. The most common preprocessing algorithms include smoothing (Gaussian blurring), sharpening, gamma correction, statistical image filtering (median, min, max), frequency band filtering, and matrix transformations for rotation/translation/skew adjustment [15].

4.9.4.2   Difference Imaging

Once preprocessing removes as much noise as possible from the input image, the input image should closely match images of the golden sample, to the point where only defects or modifications will stand out. Thus, creating a difference image between the golden sample and image under test will reveal these changes. Such a difference image is found by performing a pixel-wise subtraction:

$$I_d = I_r - I_t, \tag{4.1}$$

where I_r is the golden reference image, I_t is the image under test, and I_d is the resulting difference image. Often, only magnitude information from I_d is used, since it is the most important characteristic of a difference image. That is,

$$I_d = |I_r - I_t|. \tag{4.2}$$

Depending on the application, I_d may be further cleaned before proceeding to thresholding. If so, many of the same steps employed in preprocessing can be used.

4.9.4.3   Thresholding

Once I_d is obtained as explained above, it undergoes thresholding to convert it to a binary image. This stage varies widely depending on the application and assurance objective. However, each implementation requires a threshold T such that:

$$I_b = \begin{cases} 0, & I_d(m, n) < T \\ 1, & I_d(m, n) \geq T, \end{cases} \tag{4.3}$$

where I_b is the binarized output of this process and (m, n) represents the pixel coordinates within I_d.

While T can be a single number, it is often advantageous to employ a more complex thresholding process known as adaptive thresholding. In this scenario, the threshold is per-pixel rather than a global value (i.e., $I_d(m, n)$ is compared against $T(m, n)$). This is helpful when the threshold for a given pixel is dependent on neighboring pixel values.

As in the previous stage, I_b is usually further refined before proceeding to fault detection and localization. However, unlike the techniques used in preprocessing, I_b is cleaned with discrete rather than continuous operators to retain binary values. Noise removal methods in this category include morphological operations, Hough filtering, size filtering, and more.

4.9.4.4   Fault Detection/Localization

By this stage in the CV/IP hardware assurance process, the extracted I_b represents a black-and-white image where differences between the golden sample and board under test are highlighted. Depending on the size of differences shown in I_b, a fault may be declared at specific (m, n) coordinates within the image. In the ideal case for a non-defective sample, I_b will be a completely black image—indicating there were no significant differences between I_t and I_r.

4.9.4.5   SME Verification

Even in automated workflows, it is still common to involve a technician when confirming the existence of hardware faults. Depending on the volume of samples being tested, the SME may check every PCB or just a representative selection from each batch of fault detections.

4.10   Imaging Modalities for HW Assurance

The electromagnetic spectrum is composed of a wide variety of frequencies, most of which can be exploited by imaging devices. Longer frequencies, such as radar bands, however, do not allow for the resolution necessary to provide PCB hardware assurance. As such, only relatively short wavelengths of the electromagnetic spectrum can be used in this regard. Commonly leveraged bandwidths are shown in Fig. 4.7.

Imaging can be achieved through means other than photon wavelengths in the electromagnetic spectrum. Depending on the method used, more material properties can be obtained. Figure 4.8 shows typical images obtained with each modality discussed below, organized by the way in which they characterize the respective imaging sample.

4.10.1   Terahertz Imaging

The longest wavelength usable for PCB imaging is the terahertz (THz) region, which offers resolution from 1 mm to 10 μm. Since THz waves are strongly attenuated in atmospheric conditions, long-range imaging in this regime is extremely difficult. Hence, the imaging apparatus must be close to the sample when acquiring data [21].

Capabilities   THz imaging provides accurate depth information up to several millimeters, depending on the material being imaged. This makes 3D surface characterization of PCB components possible, which aids in detecting counterfeit

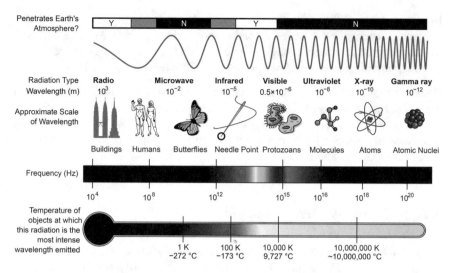

Fig. 4.7 Depiction of the relative wavelength sizes of the electromagnetic spectrum. Importantly, only those with resolution on the order of micrometers to nanometers can be used for PCB assurance. (Retrieved from Wikimedia Commons under CC license (Inductiveload, NASA, EM Spectrum Properties edit, CC BY-SA 3.0))

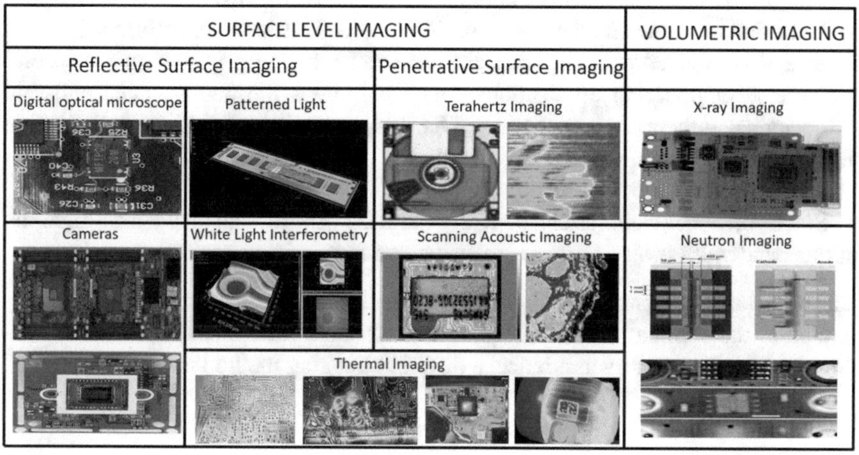

SURFACE LEVEL IMAGING			VOLUMETRIC IMAGING
Reflective Surface Imaging		Penetrative Surface Imaging	
Digital optical microscope	Patterned Light	Terahertz Imaging	X-ray Imaging
Cameras	White Light Interferometry	Scanning Acoustic Imaging	Neutron Imaging
	Thermal Imaging		

Fig. 4.8 Imaging techniques available in different electromagnetic bandwidths. In the following subsections, the primary advantages and disadvantages of each are discussed

measures, such as black-topping, polishing, remarking, and more. While it cannot penetrate through metals, it is translucent to plastics, allowing non-destructive PCB imaging through most packaging material [3].

Additionally, THz imaging provides useful information for sample material analysis. When a sample is imaged, a portion of the source beam will pass through the object, some will be reflected back toward the source, and some will be absorbed by either the sample or surrounding medium. By analyzing the reflected and transmitted energy levels, material conductance and transmittance can be obtained. Moreover, with a strong enough source, such images can be collected within minutes, providing relatively fast assessments of a sample PCB [2, 3].

Disadvantages Since terahertz imaging is still relatively new, the costs associated with THz sources are quite large. For large-bandwidth, high-power THz emission, the necessary beam source costs anywhere from \$50 to 150 K. Moreover, without sophisticated equipment, the maximum attainable resolution is significantly lower than that of other imaging bandwidths.

4.10.2 Infrared/Thermal Imaging

Infrared (IR) imaging occupies the next-lowest bandwidth of the electromagnetic spectrum after the THz regime. Offering resolution from $10 \,\mu m$ to $700 \,nm$, it can resolve distances with more accuracy than THz sources [24].

Capabilities Like the THz regime, thermal imaging also provides surface information about the sample along with limited volumetric capabilities. However, instead of characterizing material composition, the IR bandwidth is typically used to determine run-time characteristics of the sample. This includes information, such as data flow and component heat dissipation. IR imaging is particularly useful for reverse engineering applications, since it bridges the gap between software commands and their implementation in hardware. Additionally, its heat dissipation analysis allows manufacturers to perform quality assurance checks for various on-board components [23, 24].

Since consumer cameras can be easily outfitted with IR-sensitive lenses, imaging in this bandwidth is also fairly inexpensive and fast.

Disadvantages Infrared imaging is less versatile than other modalities, mostly limited to the applications discussed above. As such, for applications outside data flow analysis and quality assurance, a different imaging method will be required.

4.10.3 Optical/Visual Imaging

Optical imaging can reveal sample features as small as $380 \,nm$, which is sufficient for resolving most surface-level features of a PCB.

Capabilities Extremely cheap, portable, and fast, visual imaging is the standard method for detection most surface-mount defects or tamper marks. Numerous methods of capturing and analyzing visual light have been examined. However, two of the most common methods used for hardware assurance include digital cameras and optical microscopes [30].

Standard CCD DSLR digital cameras are capable of rapid imaging for applications not dependent on high resolution. While capable of over 10x zoom in most cases, fundamental light collection and lens limits prevent them from reaching the theoretical maximum optical resolution. Such imaging devices are often used for quality assurance when a golden sample is present and defects are large. This would include bad solder joints, broken IC pins, missing trace or silkscreen elements, and similar defects. When the features of interest are a few microns or smaller, an optical microscope is required. For large PCBs, microscope images are taken of each portion of the sample and stitched together in post-processing [25, 30].

Disadvantages Optical imaging is purely surface-level; that is, it does not provide volumetric information about the sample. Subsequently, it cannot provide information about defects or alterations underneath a mounted component or within its package.

4.10.4   X-Ray Imaging

X-ray wavelengths range from 10 to 0.1 nm. However, limitations to the CCD detector which captures and records X-ray beam locations mean the total X-ray imaging system has a resolution down to $\approx 1\,\mu$m.

Capabilities X-ray imaging is by far the widest-used method for volumetric quality assurance and tamper detection. Capable of imaging through most packaging materials and low-density metals, X-ray imaging devices can reconstruct nearly every aspect of a 3D sample as needed. For this reason, they are used for an extremely wide array of applications from BGA and TSV inspection to netlist reconstruction [5, 6, 31].

Disadvantages High-quality 3D X-ray scans require several hours to create. Hence, full volumetric reconstruction for assurance can only be performed on a small subset of PCB samples. Subsequently, this is not feasible for verification in a high-volume production scenario. Also, X-ray imaging devices with sufficient resolution are quite expensive, costing over $50 K for an entry-level device with the necessary tolerances and resolution.

4.10.5   Scanning Acoustic Imaging

Helpful information about a PCB sample can be gathered through analyzing its vibrational modes through different forms of excitation. During acoustic imaging, an ultrasonic transducer is activated at frequencies ranging from 15 to 250 MHz [37]. The resulting pressure waves collide with the sample being evaluated, causing it to vibrate in response. Depending on the quality of the source transducer, spatial resolution up to 5 μm can be obtained [10].

Capabilities Fractured PCBs and on-board components with fissures have radically different vibrational modes than their structurally sound counterparts. As such, the data retrieved from scanning acoustic imaging starkly shows defects or modifications of this nature (e.g., voids, delaminations, etc.). This modality is cheap and offers unusually high contrast for material defects as shown above, leading to its widespread use by the hardware assurance community [10].

Disadvantages While SAM is highly effective at characterizing material defects at the chip level and PCB level, multilayered boards present a challenging imaging scenario. Such samples cause a wide variety of excited acoustic modes to propagate and interfere, resulting in a complex waveform at the receiver end. Consequently, SNR is reduced and material properties are difficult to interpret [46].

4.10.6   Neutron Imaging

Only recently available, neutron imaging works using a beam of neutrons fired at the sample under test. Though the fundamental limits of neutron imaging allow for much greater spatial resolution, current measuring techniques are only capable of values greater than 5 μm [29, 44].

Capabilities By measuring the properties of these neutrons as they interact with the sample, depth and material information can be retrieved. This is possible in both 2D and 3D capacities, allowing volumetric data analysis similar to X-ray imaging. However, neutron imaging is capable of penetrating materials more dense than X-rays—provided the source beam strength is sufficient [4].

Disadvantages Also similar to X-ray systems, neutron imaging setups are slow, are expensive, and require a high degree of maintenance. Notably the fundamental limitations behind neutron imaging also inhibit this modality from passing through many kinds of plastics [4]. As a result, PCB components packaged in lightweight plastic material cannot be non-destructively analyzed with neutron imaging.

References

1. Acharya, T., Ray, A.K.: Image Processing: Principles and Applications. Wiley (2005)
2. Ahi, K., Asadizanjani, N., Shahbazmohamadi, S., Tehranipoor, M., Anwar, M.: Terahertz characterization of electronic components and comparison of terahertz imaging with x-ray imaging techniques. In: Terahertz Physics, Devices, and Systems IX: Advanced Applications in Industry and Defense, vol. 9483, p. 94830K. International Society for Optics and Photonics (2015). https://doi.org/10.1117/12.2183128

3. Ahi, K., Shahbazmohamadi, S., Asadizanjani, N.: Quality control and authentication of packaged integrated circuits using enhanced-spatial-resolution terahertz time-domain spectroscopy and imaging. Opt. Lasers Eng. **104**, 274–284 (2018). https://doi.org/10.1016/j.optlaseng.2017.07.007

4. Anderson, I.S., McGreevy, R.L., Bilheux, H.Z.: Neutron imaging and applications, vol. 2209, p. 987. Springer Science+ Business Media, LLC (2009)

5. Asadizanjani, N., Shahbazmohamadi, S., Tehranipoor, M., Forte, D.: Non-destructive pcb reverse engineering using x-ray micro computed tomography. In: 41st International Symposium for Testing and Failure Analysis, pp. 1–5. ASM (2015)

6. Asadizanjani, N., Tehranipoor, M., Forte, D.: Pcb reverse engineering using nondestructive x-ray tomography and advanced image processing. IEEE Trans. Compon. Packag. Manuf. Technol. **7**(2), 292–299 (2017). https://doi.org/10.1109/TCPMT.2016.2642824

7. Azhagan, M., Mehta, D., Lu, H., Agrawal, S., Tehranipoor, M., Woodard, D.L., Asadizanjani, N., Chawla, P.: A review on automatic bill of material generation and visual inspection on pcbs. In: ISTFA 2019: Proceedings of the 45th International Symposium for Testing and Failure Analysis, p. 256. ASM International (2019)

8. Bernard, D., Bryant, K.: What 2D and 3D (ct) x-ray inspection now provides for electronics in automotive environments. In: 2018 International Conference on Electronics Packaging and iMAPS All Asia Conference (ICEP-IAAC), pp. G–1–G–5 (2018)

9. Botero, U.J., Wilson, R., Lu, H., Rahman, M.T., Mallaiyan, M.A., Ganji, F., Asadizanjani, N., Tehranipoor, M.M., Woodard, D.L., Forte, D.: Hardware trust and assurance through reverse engineering: a survey and outlook from image analysis and machine learning perspectives. arXiv preprint: 2002.04210 (2020)

10. Brand, S., Raum, K., Czuratis, P., Hoffrogge, P.: Signal analysis in scanning acoustic microscopy for non-destructive assessment of connective defects in flip-chip bga devices. In: 2007 IEEE Ultrasonics Symposium Proceedings, pp. 817–820. IEEE (2007). https://doi.org/10.1109/ULTSYM.2007.209. http://ieeexplore.ieee.org/document/4409782/

11. Carlsson, J.: Crosstalk on printed circuit boards. 2 (1994)

12. Chen, H.C., Lin, S.C.: The study of using x-ray laminography on printed-circuit board inspection. In: Hinduja, S., Fan, K.C. (eds.) Proceedings of the 35th International MATADOR Conference, pp. 221–224. Springer, London (2007)

13. Chen, Z., Yang, M., Shi, Q., Kuang, X., Qi, H.J., Wang, T.: Recycling waste circuit board efficiently and environmentally friendly through small-molecule assisted dissolution. Sci. Rep. **9**(1), 17902 (2019). https://doi.org/10.1038/s41598-019-54045-w

14. Ghosh, S., Basak, A., Bhunia, S.: How secure are printed circuit boards against trojan attacks? IEEE Des. Test **32**(2), 7–16 (2015)

15. Gonzalez, R.C., Woods, R.E., Masters, B.R.: Digital image processing, 3rd ed. J. Biomed. Opt. **14**(2), 029901 (2009). https://doi.org/10.1117/1.3115362

16. Guin, U., DiMase, D., Tehranipoor, M.: Counterfeit integrated circuits: detection, avoidance, and the challenges ahead. J. Electron. Test. **30**(1), 9–23 (2014)

17. Harrison, J.: System-level hardware Trojan taxonomy (2020). https://trust-hub.org/benchmarks/system-level-trojan

18. Joong Hong, J., Ja Park, K., Gu Kim, K.: Parallel processing machine vision system for bare pcb inspection. IECON '98. In: Proceedings of the 24th Annual Conference of the IEEE Industrial Electronics Society (Cat. No.98CH36200) vol.3, pp. 1346–1350 (1998)

19. Hong, X., Cheng, D., Shi, Y., Lin, T., Gwee, B.H.: Deep learning for automatic IC image analysis. In: 2018 IEEE 23rd International Conference on Digital Signal Processing (DSP), pp. 1–5 (2018). https://doi.org/10.1109/ICDSP.2018.8631555

20. Hu, T., Wu, L., Zhang, X., Yin, Y., Yang, Y.: Hardware trojan detection combine with machine learning: an svm-based detection approach. In: 2019 IEEE 13th International Conference on Anti-counterfeiting, Security, and Identification (ASID), pp. 202–206. IEEE (2019). https://doi.org/10.1109/ICASID.2019.8924992. https://ieeexplore.ieee.org/document/8924992/

21. Kaltenecker, K.J., Kelleher, E.J.R., Zhou, B., Jepsen, P.U.: Attenuation of thz beams: a "how to" tutorial. J. Infrared Millimeter Terahertz Waves **40**(8), 878–904 (2019). https://doi.org/10.1007/s10762-019-00608-x

22. Kim, H.W., Yoo, S.I.: Defect detection using feature point matching for non-repetitive patterned images. Pattern. Anal. Applic. **17**(2), 415–429 (2014). https://doi.org/10.1007/s10044-012-0305-7

23. Maldaque: Infrared Methodology and Technology. CRC Press (1994). Google-Books-ID: xPOpMYKCTBgC

24. Malyutenko, V.K.: High resolution infrared "vision" of dynamic electron processes in semiconductor devices (abstract). Rev. Sci. Instrum. **74**(1), 655–655 (2003). https://doi.org/10.1063/1.1521527

25. Mar, N.S.S., Yarlagadda, P.K.D.V., Fookes, C.: Design and development of automatic visual inspection system for pcb manufacturing. Robot. Comput. Integr. Manuf. **27**(5), 949–962 (2011). https://doi.org/10.1016/j.rcim.2011.03.007

26. Matlin, E., Agrawal, M., Stoker, D.: Non-invasive recognition of poorly resolved integrated circuit elements. IEEE Trans. Inf. Forensics Secur. **9**(3), 354–363 (2014). https://doi.org/10.1109/TIFS.2013.2297518. Conference Name: IEEE Transactions on Information Forensics and Security

27. Milkovich, C.: Assembly of Area Array Components, pp. 762–803. Springer US, Boston (2001). https://doi.org/10.1007/978-1-4615-1389-6_19

28. Millennium Circuits Limited: PCB Testing Methods Guide. Technical report (2019). https://www.mclpcb.com/pcb-testing-methods-guide/

29. Minniti, T., Tremsin, A.S., Vitucci, G., Kockelmann, W.: Towards high-resolution neutron imaging on imat. J. Instrum. **13**(01), C01039–C01039 (2018). https://doi.org/10.1088/1748-0221/13/01/C01039

30. Moganti, M., Ercal, F., Dagli, C.H., Tsunekawa, S.: Automatic pcb inspection algorithms: a survey. Comput. Vis. Image Underst. **63**(2), 287–313 (1996). https://doi.org/10.1006/cviu.1996.0020

31. Neubauer, C.: Intelligent x-ray inspection for quality control of solder joints. IEEE Trans. Compon. Packag. Manuf. Technol. Part C **20**(2), 111–120 (1997). https://doi.org/10.1109/3476.622881

32. Pau, L.F.: Computer Vision for Electronics Manufacturing. Springer Science & Business Media (2012). Google-Books-ID: VJbaBwAAQBAJ

33. Quadir, S.E., Chen, J., Forte, D., Asadizanjani, N., Shahbazmohamadi, S., Wang, L., Chandy, J., Tehranipoor, M.: A survey on chip to system reverse engineering. ACM J. Emerg. Technol. Comput. Syst. (JETC) **13**(1), 1–34 (2016)

34. Rahman, M.T., Shi, Q., Tajik, S., Shen, H., Woodard, D.L., Tehranipoor, M., Asadizanjani, N.: Physical inspection & attacks: new frontier in hardware security. In: 2018 IEEE 3rd International Verification and Security Workshop (IVSW), pp. 93–102. IEEE (2018)

35. Robertson, J., Riley, M.: The big hack: how China used a tiny chip to infiltrate U.S. companies. Technical report (2018)

36. Rosenfeld, K., Karri, R.: Attacks and defenses for jtag. IEEE Des. Test Comput. **27**(1), 36–47 (2010)

37. Semmens, J.E., Kessler, L.W.: Application of acoustic frequency domain imaging for the evaluation of advanced micro electronic packages. Microelectron. Reliab. **42**(9–11), 1735–1740 (2002). https://doi.org/10.1016/S0026-2714(02)00222-6

38. Shih, F.Y.: Image processing and pattern recognition: fundamentals and techniques. IEEE Press/Wiley (2010)

39. Sood, B., Pecht, M.: Controlling moisture in printed circuit boards. IPC Apex EXPO Proceedings (2010)

40. Tehranipoor, M.M., Guin, U., Forte, D.: Counterfeit integrated circuits. In: Counterfeit Integrated Circuits, pp. 15–36. Springer (2015)

41. Torrance, R., James, D.: The state-of-the-art in ic reverse engineering. In: International Workshop on Cryptographic Hardware and Embedded Systems, pp. 363–381. Springer (2009)

42. Vishwakarma, G., Lee, W.: Exploiting jtag and its mitigation in iot: a survey. Futur. Intern. **10**(12), 121 (2018). https://doi.org/10.3390/fi10120121

43. Wang, W.C., Chen, S.L., Chen, L.B., Chang, W.J.: A machine vision based automatic optical inspection system for measuring drilling quality of printed circuit boards. IEEE Access **5**, 10817–10833 (2017). https://doi.org/10.1109/ACCESS.2016.2631658

44. Williams, S.H., Hilger, A., Kardjilov, N., Manke, I., Strobl, M., Douissard, P.A., Martin, T., Riesemeier, H., Banhart, J.: Detection system for microimaging with neutrons. J. Instrum. **7**(02), P02014–P02014 (2012). https://doi.org/10.1088/1748-0221/7/02/P02014

45. Xu, X., Rahman, F., Shakya, B., Vassilev, A., Forte, D., Tehranipoo, M.: Electronics supply chain integrity enabled by blockchain. ACM Trans. Des. Autom. Electron. Syst. **24**(3), 10.1145/3315571 (2019). https://doi.org/10.1145/3315571. https://pubmed.ncbi.nlm.nih.gov/32116465. 32116465[pmid]

46. Yu, Z., Boseck, S.: Scanning acoustic microscopy and its applications to material characterization. Rev. Mod. Phys. **67**(4), 863–891 (1995). https://doi.org/10.1103/RevModPhys.67.863

Chapter 5
Electrical Probing Attacks

5.1 Introduction

Physical attacks are capable of bypassing the confidentiality and integrity provided by modern cryptography through observation of a chip's silicon implementation. Such attacks are especially threatening to the integrated circuits (ICs) in smart-cards, smartphones, military systems, and financial systems that process sensitive information. Unlike non-invasive side-channel analysis (e.g., power or timing analysis), probing contacts the internal wires of a security-critical module to extract information. Probing, like reverse engineering and circuit edit, poses a grave threat to mission-critical applications and thus demands the development of effective countermeasures from the research community [10, 15–17].

Probing attacks are already a threat. The most recent example of a probing attack emerged when FBI requested help in defeating the passcode retry counter of the Apple iPhone 5c owned by a terrorist suspect. Researchers reverse engineered the proprietary protocol used by the phone's NAND flash, copied the contents, and then brute forced the passcode in less than a day [12]. While in this case the attack was conducted by researchers, compromise of military technologies through probing could have catastrophic consequences that cost lives. In such instances, advanced IC failure analysis and debug tools are used to probe the ICs internally. Among such tools, a focused ion beam (FIB) is the most dangerous. FIBs use ions at high beam currents for site-specific milling and material removal. The same ions can also be injected close to a surface for material deposition. These capabilities allow FIBs to cut or add traces to the substrate within a chip, enabling them to redirect signals, modify trace paths, and add/remove circuits. Though FIB was initially designed for failure analysis, a skilled attacker can use it to obtain on-chip keys, establish privileged access to memory, obtain device configuration, or inject faults. This can be accomplished by rerouting them to an existing output pin, creating a new contact for probing, or re-enabling IC test mode. Most of these techniques would not be possible without a FIB. While countermeasures against probing such

© The Author(s), under exclusive license to Springer Nature Switzerland AG 2021
N. Asadizanjani et al., *Physical Assurance*, https://doi.org/10.1007/978-3-030-62609-9_5

as active meshes, optical sensors, and analog sensors have been proposed, they are clumsy, expensive, and ad hoc. It has been shown repeatedly that an experienced FIB operator can easily bypass them via circuit edit. In [13], well-known hacker Christopher Tarnovsky probed the firmware of the Infineon SLE 66CX680P/PE security/smart chip from the frontside (i.e., top metal layer) by rewiring its active mesh and making contact with its buses using FIB.

We expect the growth of FIB-assisted probing attacks for a variety of reasons. FIBs are becoming cheaper and easier to access than ever before (e.g., FIB time can be purchased for a couple hundred dollars per hour). Further, as FIB capabilities continue to improve for failure analysis, more powerful attacks will be enabled. In contrast, non-invasive and semi-invasive attacks either do not scale to modern semiconductors with Moore's law or can be mitigated by inexpensive countermeasures. As non-invasive and semi-invasive attacks continue to become less effective, one can expect attackers to migrate to FIB. For these reasons, it is of the utmost importance that we stay ahead of attackers and develop more effective countermeasures against FIB-based probing. Since FIB capabilities are almost limitless, the best approaches should make probing as costly, time-consuming, and frustrating as possible. A significant challenge in doing so lies in the fact that the time, effort, and cost to design a FIB-resistant chip must remain reasonable, especially to design engineers who are generally not security experts. This could be especially important in the upcoming Internet-of-Things (IoT) era which will likely consist of an abundance of low-end chips that are easily physically accessed.

In this section, we present state-of-the-art research in the field of circuit edit and anti-probing, highlight the challenges, and offer future research directions for CAD and test communities. The rest of the paper is organized as follows. Section 5.2 reviews technical background related to probing attacks. The steps of a typical probing attack are illustrated in Sect. 5.3. Section 5.4 introduces existing countermeasures against probing attacks and their limitations. At last, we conclude the chapter in Sect. 5.6.

5.2 Probing Attack Fundamentals

Comprehension of the adversary's goal and the techniques he/she uses to success-fully carry out probing is the first step in overcoming this significant threat. In this section, we review technical details of the probing process and make associations between technical requirements, decisions, and perceived limitations of state-of-the-art techniques.

5.2.1 Probing Attack Targets

It is essential for both attackers and countermeasure designers to determine which signals are more likely to be targeted in a probing attack. We term such signals as *assets*. An asset is a resource of value which is worth protecting from an adversary [1]. Unfortunately, a more palpable definition of asset has not been proposed or agreed upon. To help illustrate the wide range of possible information that could be assets, here we enumerate a few quintessential examples of assets that are the most likely targets for probing attacks.

Keys Keys of an encryption module (e.g., private key of a public key algorithm) are archetypal assets. They are usually stored in non-volatile memory on the chip. If the key is leaked, the root of trust it provides will become compromised and could serve as a gateway to more serious attacks. An example is original equipment manufacturer (OEM) keys that are used to grant legitimate access to a product or chip. Leakage of such keys will result in tremendous loss of revenue for the product owner, denial of service, or information leakage.

Firmware and Configuration Bitstream Electronic intellectual properties (IPs) such as low-level program instruction sets, manufacturer firmware, and FPGA configuration bitstreams are often sensitive, are mission-critical, and/or contain trade secrets of the IP owner. Once compromised, counterfeiting, cloning, or exploits of system vulnerabilities could be facilitated.

On-Device Protected Data Sensitive data, such as health and personal identifiable information, should be kept private. Leakage of such information could result in fraud, embarrassment, or property/brand damage for the data owner.

Device Configuration Device configuration data control the access permissions to the device. They specify which services or resources can be accessed by each individual user. If the configurations are tampered with, an attacker could illegally gain access to resources denied to him otherwise.

Cryptographic Random Number Hardware-generated random numbers, such as keys, nonces, one-time pads, and initialization vectors for cryptographic primitives, also require protection. Compromising this type of asset will weaken the cryptographic strength of the digital services on the device.

5.2.2 Essential Technologies of a Probing Attack

A successful probing attack entails a time-consuming and sophisticated process. Countermeasure designers are often interested in ways to make this process go astray. For this purpose, we examine the central approaches and technologies used in published attacks in the following subsections.

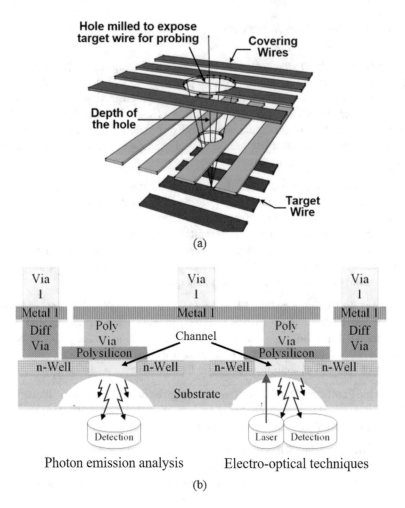

Fig. 5.1 (**a**) Milling from BEOL through covering wires (purple and green) to reach target wires (blue). (**b**) Optical probing: photon emission (PE) and electro-optical frequency modulation (EOFM) or laser voltage techniques (LVX) are used for passive and active measurements, respectively

Frontside vs. Backside Probing attack targets are those metal wires that carry assets, henceforth called target wires. The most common approach to reach target wires is to expose them from the back end of line (BEOL), i.e., from the top metal layer toward silicon substrate (illustrated in Fig. 5.1a). This is called a frontside probing attack. Exposure of target wires is first facilitated with FIB milling, and then an electric connection to the target wire can be established, e.g., by conductor deposition capability of the FIB. Finally, extraction of sensitive information ensues.

A backside probing attack, i.e., probing that occurs through the silicon substrate, was proposed in [2]. Backside attack targets are not limited to wires. By exploiting a phenomenon during transistor activity known as photon emission, transistors can also be probed to extract information.

Electrical Probing vs. Optical Probing The method to access assets shown in Fig. 5.1a is typical for electrical probing, i.e., accessing an asset carrying signal via electrical connection. A different approach is optical probing as shown in Fig. 5.1b. Optical probing techniques are often used in backside probing to capture photon emission phenomena during transistor switching. When transistors are switching, they spontaneously emit photons without external stimuli. By passively receiving and analyzing the photons emitted from a specific transistor, the signal processed by that transistor can be inferred. Compared to electrical probing, the optical approach has the advantage of being a purely passive observation, which makes it very difficult to detect. In addition to photon emission analysis, laser voltage technique (LVX) and electro-optical frequency modulation (EOFM) are also used during backside attacks. These techniques actively illuminate the switching transistors and then infer asset signal values by observing the reflected light.

The primary deficiency of optical probing lies in the fact that photons emitted in these techniques are infrared due to silicon energy band gap, which has a wavelength of 900 nm or higher [2]. Therefore, the optical resolution between transistors is limited to within one order of magnitude of the wavelength due to Rayleigh criterion.

5.3   Essential Steps of a Probing Attack

In this subsection, we continue our examination of probing attack fundamentals by outlining its essential steps.

Decapsulation The first stage of most invasive physical attacks is to either partially or fully remove the chip package in order to expose the silicon die. This requires adequate practice and expertise in handling harmful chemicals. Acid solutions such as fuming nitric acid combined with acetone at 60 °C are often used to remove plastic packages [11]. Decapsulation can also be done from the backside of the chip by removing the copper plate mechanically without chemical etching.

Reverse Engineering Reverse engineering [8] is the process of extracting design information from something, typically to reproduce it. In the case of probing, reverse engineering is used to understand how the chip works, which requires that the layout and netlist be extracted. By studying the netlist, the attacker can identify the assets. One-to-one correspondence between the netlist and layout can then determine the locations of target wires and buses and, in the event where cutting off a wire is unavoidable, determine whether the cut would impact asset extraction. State-of-the-art tools such as ICWorks from Chipworks can perform automatic extraction of netlists from images of each layer taken with optical or scanning electron microscopes (SEM in Fig. 5.2a), which greatly reduces the attacker's effort.

(a)   (b)

Fig. 5.2 (**a**) Scanning electron microscope (SEM). (**b**) Focused ion beam (FIB). Note that the attacker does not need to purchase all these instruments since rent by time is quite low cost

Locating Target Wires Once the probing wire targets have been identified by reverse engineering, the next stage is locating the wires associated with the target on the IC under attack. The crux of the problem here is that while the attacker has located target wires on sacrificial devices during reverse engineering process, he/she now has to find the absolute coordinates of the point to mill blindly. This requires a precise-enough kinematic mount and fiducial markers (i.e., visual points of reference on the device) to base these absolute coordinates.

Reaching Target Wire and Extracting Information With the help of modern circuit editing tools like FIB (see Fig. 5.2b), a hole can be milled to expose the target wire. State-of-the-art FIBs can remove and deposit material with nanometer resolution, which allows an attacker with a FIB to edit out obstructing circuitry or deposit conducting paths that may serve as electrical probe contacts. This feature indicates that many countermeasures can be disabled by simply disconnecting a few wires and that a FIB-equipped attacker could field as many concurrent probes as logic analyzer allows. Once a target wire is exposed and assuming it is contacted without triggering any probing alarm signals from active or analog shields, the asset signals need to be extracted, for example, with a probe station. The difficulty of this step depends on a few factors. First, software and hardware processes might need to be completed before the asset is available. Further, the sensitive information may not be in the same clock cycle; if the chip has an internal clock source to prevent external manipulation, the attacker will need to either disable it or synchronize his own clock with it.

5.4 Existing Countermeasures and Limitations

In the past decade, researchers have proposed various technologies to protect security-critical circuits against probing attacks. In this section, we review a few representative countermeasures and highlight their limitations. Unfortunately, to date, none of them offer a satisfactory solution. Further, to the best of our knowledge, no method has been proposed to adequately address backside probing attacks.

5.4.1 Active Shields

Active shield is so far the most investigated probing countermeasure. In this approach, a shield which carries signals is placed on the top-most metal layer to detect holes milled by FIB. The shield is referred to as "active" because signals on these top layer wires are constantly monitored to detect if milling has cut them [3]. Figure 5.3 shows one illustrative example. As shown in the figure, a digital pattern is generated from a pattern generator, transmitted through the shield wires on top-most metal layer, and then compared with a copy of itself transmitted from lower layer. If an attacker mills through the shield wires on top layer to reach target wire, the hole is expected to cut open one or more shield wires, thereby leading to a mismatch at

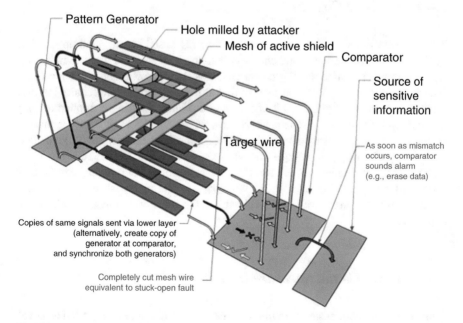

Fig. 5.3 Basic working principle of active shields

the comparator and triggering an alarm signal to erase or stop generating sensitive information. Despite its popularity, active shields are not without shortcomings. Their biggest problems are that they impose large overheads on the design, but at the same time are very vulnerable to attacks with advanced FIBs, e.g., circuit editing attacks.

5.4.2 Analog Shields and Sensors

An alternative approach to active shield is to construct an analog shield. Instead of generating, transmitting, and comparing digital patterns, analog shields monitor parametric disturbances with its mesh wires.

In addition to shield designs, the probe attempt detector (PAD) [6] (as shown in Fig. 5.4) also uses capacitance measurement on selected security-critical wires to detect additional capacitance introduced by a metal probe. Compared to active shields, analog shields detect probing without test patterns and require less area overhead. The PAD technique is also unique in remaining effective against electrical probing from the backside. The problem with analog sensors or shields is that analog measurements are less reliable due to process variations, a problem further exacerbated by feature scaling.

5.4.3 t-Private Circuits

The t-private circuit technique is proposed in [5] based on the assumption that the number of concurrent probe channels that an attacker could use is limited and exhausting this resource thereby deters an attack. In this technique, the circuit of a security-critical block is transformed so that at least t+1 probes are required within one clock cycle to extract one bit of information. First, masking is applied to split computation into multiple separate variables, where an important binary signal, x, is encoded into t+1 binary signals by XORing it with t independently generated random signals ($r_{(t+1)} = x \oplus r_1 \oplus \ldots \oplus r_t$) as shown in Fig. 5.5. Then, computations on x are performed in its encoded form in the transformed circuit. x can be recovered (decoded) by computing $x = r_1 \oplus \ldots \oplus r_t \oplus r_{(t+1)}$. The major issue with t-private circuit is that the area overhead involved for the transformation is prohibitively expensive.

5.4.4 Other Countermeasure Designs

Some other countermeasures are implemented in real ICs but less reported as novel designs because they are more or less dated. One known countermeasure that deters

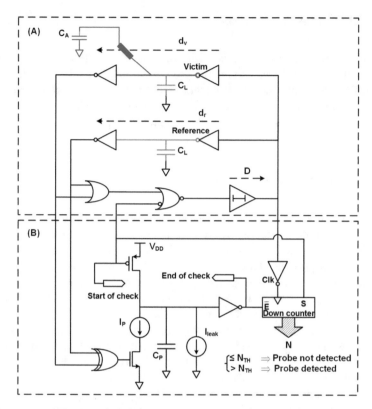

Fig. 5.4 Probing attempt detector (PAD)

Fig. 5.5 Input encoder (left) and output decoder (right) for masking in t-private circuits

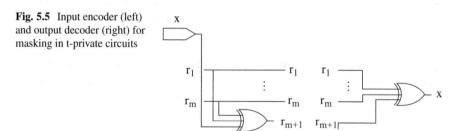

decapsulation stage of probing attacks is light sensor that is sometimes included in a tamper-resistant design. Some other techniques include scrambling wires and avoiding repetitive patterns in shield mesh to impede the locating-target-wire stage of probing attacks. They are not particularly effective as exploits against them have been detailed in [13].

5.5 CAD for Anti-probing

Existing countermeasures are ad hoc with inefficient protection, are not designed to counter FIB-based attack, and require prohibitive area and design overhead [15]. Further, there is no holistic and efficient approach that can be easily incorporated into conventional application-specific integrated circuit (ASIC) design flow to protect security-critical circuits and nets from probing attack.

5.5.1 Anti-probing Design Flow

The objective is to develop a FIB-aware anti-probing physical design flow that incorporates automated security-aware floor-planning, cell placement, routing, and evaluation into the conventional flow in order to protect the security-critical nets against frontside probing attacks. We shall accomplish this by using a chip's internal functional nets as "shield" nets on upper layers to provide coverage for "target" nets (i.e., those carrying asset signals) on lower layers in the design. Another copy of shield nets will be routed in lower layers. Once at least one shield net on upper layer is cut off in an attack, the comparator will detect the mismatch between the signal on upper shield net and the one from lower layer. An alarm will be triggered to take the appropriate actions (e.g., terminate the operation of the chip or remove all asset information). Note that by leveraging the internal functional nets of the design itself for protection without adding any extra large circuitry, like the pattern generator and shielding circuit in active shield approach, the overhead of our approach is very low. In addition, when shield nets are placed within internal metal layers, they will be far more difficult for an attacker to bypass and reroute than dedicated shields, like active shield which typically resides at top metal layer, since the metal wires above the shield layer will be a huge obstacle to circumvent during the attack. Also, our approach can complement those PAD-based techniques, e.g., PAD can protect buses efficiently, while our approach can protect PAD and other non-bus circuits. Further, the whole anti-probing physical design flow is implemented using computer-aided design (CAD) tools, which means the whole process could be completely automatic and uniform for different designs so that the design overhead to build the proposed internal shield will be very limited.

The overall workflow of our anti-probing physical design flow is shown in Fig. 5.6. It brings into conventional ASIC design flow three new steps. First, appropriate shield nets and target nets are identified for optimal protection against probing attack. Sections 5.5.1.1 and 5.5.1.2 will illustrate the detailed requirements and metrics to identify target nets and shield nets, respectively. User input includes asset information and threshold values to identify target nets and shield nets. Then, a comparator is inserted in the gate-level netlist of the original design to detect the mismatch. The comparator itself is also protected as potential probing target. The length of the comparator is determined by the number of shield nets needed

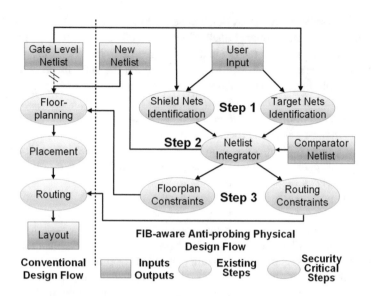

Fig. 5.6 Overall FIB-aware anti-probing physical design flow

for the dedicated design. Both inputs of the comparator are connected to the same source nets, but one is the exact shield net from upper layer, while the other one is the copy form lower layer. These will be implemented in the routing constraint step as illustrated in Sect. 5.5.1.5. Selection of the best layer for single-layer shield and the best two layers for multilayer shield will be discussed in Sect. 5.5.1.3. Next, floor-planning and wire routing of the design are constrained to build the internal shield and provide protection on target nets against probing attacks. Details of these procedures are given in Sects. 5.5.1.4 and 5.5.1.5. At last, to evaluate the protection performance of shielded designs, the exposed area metric [9] with additional realistic optimization is used as discussed in Sect. 5.5.1.6.

5.5.1.1   Target Net Identification

In this subsection, we discuss how we identify the nets which are most likely to be targeted for probing. Nets that are connected to assets are the most likely to be probed. In addition, an attacker can probe nets that are not directly connected to an asset, but still contain valuable information from which the asset can be derived. For example, let us assume that a two-input XOR gate where one input is connected to an asset, e.g., encryption key, and the other input is connected to an input that an attacker can control, e.g., plain-text. Then, the attacker can infer the asset by controlling the plain-text input to logic "0" and probing the output of the XOR gate because the asset input is consistent with the output when the other input of the XOR gate is logic "0." Therefore, in addition to nets that are directly connected to assets, other nets which can be exploited to extract the asset also need to be protected against probing attack. Since it is inefficient to protect all nets in an SoC, we develop a probing target identification metric to rank the nets according to their ability to leak asset information, and therefore, the nets' likelihood of being targeted for probing can be deduced. *Note that this target identification metric only applies for the possible information leakage from pure signal propagation and simple logic combinations. Those nets that can be used to derive asset information by complicated mathematics process, e.g., the nets in the last round of an encryption module for typical fault injection attacks, are not covered in the target identification metric. To protect this type of nets against probing attack, these nets can be declared as a kind of special assets in the user input.*

Our anti-probing design flow requires the designers to input the name of nets/ports where the asset is located (e.g., the name of key nets) as user input in the first step. Then, our flow performs the target net identification technique to identify all nets which are likely to be targeted for probing attack. This technique utilizes a *target score* ($f_{TS}(i)$) metric to quantify the likelihood of a net to be targeted in a probing attack. A higher value of *target score* indicates the net is more likely to be targeted in a probing attack. The attacker will prefer to probe those high *target score* nets because he/she can reveal more information with less effort on controlling signals from the net with higher *target score*. For each net i in the circuit:

$$f_{TS}(i) = \frac{f_{IL}(i)}{f_{PD}(i) + 1} \tag{5.1}$$

where $f_{IL}(i)$ denotes information leakage and quantifies the amount of asset information leaked by observing net i and $f_{PD}(i)$ indicates the difficulty to control the logic values of internal nodes to avoid the asset signal being muted and propagate it to net i based on the SCOAP controllability metric [4]. The +1 is to avoid 0 value at the denominator. A larger value of $f_{IL}(i)$ means more asset information can be leaked at net i. On the other hand, a larger $f_{PD}(i)$ value indicates that it is more difficult to propagate an asset signal to net i. Hence, a higher $f_{TS}(i)$ represents a higher likelihood of being targeted for probing.

$f_{IL}(i)$ Calculation Information leakage (IL) of a net i quantifies how much sensitive information can be directly inferred if this net is probed and observed by the attacker. If k (0 or 1) is observed at net i, the $f_{IL.k}(i)$ is defined as the number of asset bits that net i is associated with divided by the number of possible logic combinations of the associated asset bits to output k at net i. The overall information leakage $f_{IL}(i)$ is the weighted summation of $f_{IL.0}(i)$ and $f_{IL.1}(i)$ based on the probability of observing 0 and 1 at net i. $f_{IL}(i)$ is calculated for each net and is evaluated on a gate-by-gate basis from input to output. We use a 2-input AND gate, as shown in Fig. 5.7a, as an example to present how $f_{IL}(i)$ is derived. Note that a similar process is used to evaluate $f_{IL}(i)$ for all types of standard cell gates. We classify the information leakage calculation into the following three categories:

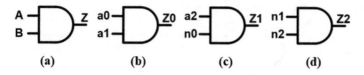

Fig. 5.7 AND gate examples

Case 1: All inputs are fan-out nets of assets In this case, all inputs of the gate are associated with the assets. Figure 5.7b shows an example of Case 1, where $a0$ and $a1$ are both asset signals. If an attacker probes the net $Z0$, then he/she can extract some information about the asset $a0$ and $a1$. We can use the following four Equations (5.2a)~(5.2d) to calculate the information leakage at $Z0$ ($f_{IL}(Z0)$):

$$f_{C,k}(Z0) = \sum_{Gate(m,n)=k} f_{C,m}(a0) \times f_{C,n}(a1) \quad (k, m, n \in \{0, 1\}) \tag{5.2a}$$

$$f_B(Z0) = f_B(a0) + f_B(a1) \tag{5.2b}$$

$$f_{IL,k}(Z0) = \frac{f_B(Z0)}{f_{C,k}(Z0)} \tag{5.2c}$$

$$f_{IL}(Z0) = \sum_{k=0}^{1} f_{IL,k}(Z0) \times \frac{f_{C,k}(Z0)}{2^{f_B(Z0)}} = \frac{f_B(Z0)}{2^{f_B(Z0)-1}} \tag{5.2d}$$

where k, m, and n are the logic value, 0 or 1, and $Gate(m,n)=k$ is the gate function to make k at the output with two inputs m and n (m AND $n = k$, in Fig. 5.7b example). Six numerical measures ($k = 0$ or 1) for input nets, e.g., $a0$ and $a1$, are considered as shown in Table 5.1. All measures for asset nets (e.g., $a0$, $a1$, and $a2$) would be 1, while they would be 0 for non-asset nets that lie outside of any asset propagation path (e.g., $n0$, $n1$, and $n2$). These measures for nets in Fig. 5.7b are shown in Table 5.2. The information leakage calculation for other types of gates is similar to AND gates. For all types of gates, Equations (5.2a)–(5.2d) are the same, while the *Gate* function needs to update accordingly. However, the total $f_{IL}(Z0)$ (Equation (5.2d)) is not a function of $f_{C,0}(Z0)$ and $f_{C,1}(Z0)$, which means the total information leakage calculation for different types of gates is a uniform function of the number of asset bits that the calculated net is associated with. Therefore, only Equation (5.2d) is needed to calculate the total information leakage for any net in the circuit. If we want to know the specific information leakage when a specific value, 0 or 1, is observed at net i, all four Equations (5.2a)~(5.2d) should be calculated.

Case 2: One of the inputs is fan-out net of assets In this case, one input of the gate is associated with assets, while the rest input is controllable by the attacker. Figure 5.7c shows an example of case 2, where $a2$ is an asset net and $n0$ is a non-asset net that is not associated with any asset but can be controlled by an attacker. Here, the attacker can control $n0$ to observe $a2$ from $Z1$. Therefore, the information

Table 5.1 Measures to calculate information leakage

Measures	Description
$f_{C,k}(i)$	Number of asset signal combinations to output k (0 or 1) at net i
$f_B(i)$	Number of asset bits in the fan-in of net i
$f_{IL,k}(i)$	Information leakage when net i is k (0 or 1)
$f_{IL}(i)$	Overall information leakage of net i

Table 5.2 Information leakage measures for nets in Fig. 5.7

Measures	a0	a1	Z0	a2	n0	Z1	n1	n2	Z2
$f_{C.0}(i)$	1	1	$1+1+1=3$	1	0	1	0	0	0
$f_{C.1}(i)$	1	1	1	1	0	1	0	0	0
$f_B(i)$	1	1	$1+1=2$	1	0	1	0	0	0
$f_{IL.0}(i)$	1	1	2/3	1	0	1	0	0	0
$f_{IL.1}(i)$	1	1	2/1	1	0	1	0	0	0
$f_{IL}(i)$	1	1	$1/2+1/2=1$	1	0	1	0	0	0

leakage for Z1 is the same as asset input a2. The information leakage measures for nets in Fig. 5.7c are shown in Table 5.2.

Case 3: No input is fan-out net of assets In this case, both inputs of the gate are non-asset signals that are not associated with any asset. Figure 5.7d shows an example of case 3, where n1 and n2 are non-asset nets. Therefore, the information leakage for Z2 is 0. The information leakage measures for nets in Fig. 5.7d are shown in Table 5.2.

Case 4: One of the inputs is a non-controllable constant value In this case, one input of the gate is associated with assets, while the rest input is a constant value which cannot be controlled by attackers. There are two scenarios that may happen. One is that the constant value will propagate the asset signal to the output of the gate, e.g., if the constant value is 1 for an AND gate. In this case, the information leakage measures in Table 5.1 of the gate output would be the same with the asset input. The other situation is that the constant value will mute the asset signal, e.g., if the constant value is 0 for an AND gate. Then, the information leakage value of the gate output would be the same with a non-asset signal which is 0.

$f_{PD}(i)$ Calculation The $f_{PD}(i)$ quantifies the difficulty to propagate asset information to net i using SCOAP combinational controllability metric (CC0 and CC1) [4]. When both inputs of a gate are fan-out nets of asset which have non-zero information leakage value (e.g., Fig. 5.7b), there is no need to control other nets to propagate asset information to the output Z0. Therefore, the $f_{PD}(Z0)$ for Z0 is set to 0. When one of the inputs is fan-out net of asset (e.g., a2 in Fig. 5.7c), to propagate a2's information to Z1, n0 needs to be 1. $CC1_{n0}$ measures the 1-controllability value for net n0. Assuming n0 is a primary input, the $CC1_{n0}$ would be 1 and $f_{PD}(Z1) = CC1_{n0} = 1$ for Z1. When net i is located n stages after asset signals, the $f_{PD}(i)$ is

Table 5.3 Target score calculation for nets in Fig. 5.7

Measures	a0	a1	Z0	a2	n0	Z1	n1	n2	Z2
$f_{IL}(i)$	1	1	1	1	0	1	0	0	0
$f_{PD}(i)$	0	0	0	0	1	1	0	0	0
$f_{TS}(i)$	1	1	1	1	0	1/2	0	0	0

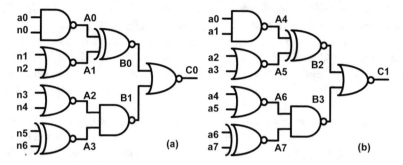

Fig. 5.8 Target score metric sample circuits

Table 5.4 Target score calculation for nets in Fig. 5.8

Net	CC0	CC1	$f_{IL}(i)$	$f_{PD}(i)$	$f_{TS}(i)$
a0–a7	INF	INF	1	0	1
n0–n6	1	1	0	0	0
A0	INF	2	1	$CC1_{n0} = 1$	0.5
A1	2	3	0	0	0
A4–A7	INF	INF	1	0	1
B0	5	5	1	$CC1_{A1} = 3$	0.25
B1	7	3	0	0	0
B2–B3	INF	INF	0.5	0	0.5
C0	4	9	1	$CC1_{A1} + CC0_{B1} = 10$	1/11
C1	INF	INF	1/16	0	1/16

the summation of n 1/0-controllability values of the non-asset input of the gate for each stage to propagate asset information to next stage.

Target Score **Calculation** Table 5.3 shows the target score calculation using Equation (5.1) for Z0, Z1, and Z2 in Fig. 5.7b, c, d, assuming n0, n1, and n2 are non-asset primary inputs. In Fig. 5.7d, since both inputs are non-asset nets without any information leakage, the *Target Score* for Z2 is 0. Figure 5.8 and Table 5.4 show the target score metric calculation on two sample circuits where different types of gates and inputs are mixed. a0–a7 are asset signals, while n0–n6 are non-asset primary inputs. In Fig. 5.8a, the information leakage value ($f_{IL}(i)$) on a0 propagation path (a0-A0-B0-C0) stays at 1, the target score is decreasing due to the difficulty to control nets (n0 = 1, A1 = 1, B1 = 0) to propagate asset information to next stage increasing. On the other hand, in Fig. 5.8b, all inputs are asset signals and the information leakage values decrease stage by stage, which indicates they are less and less likely to be targeted in probing attack.

To implement the target identification metric on a large circuit, the target score needs to be calculated from primary inputs to the primary outputs. A flip-flop can be treated as a buffer which maintains the target score and information leakage values as its input. For those feedback nets, in the initialization stage, they can be simplified as non-asset nets with zero target score and information leakage values. With the process of target score calculation, they will finally be assigned an updated value for target score and information leakage. Considering the sensitivity of the asset and the acceptable protection overhead, we can set a threshold value for target score to identify nets which are most likely targeted for probing attack. Any net whose target score is larger than this threshold value needs to be protected accordingly against probing attack. It can be observed from Fig. 5.8 and Table 5.4 that the target scores for {C0, C1} are much lower than the other nets closer to asset nets. Therefore, we may exclude {C0, C1} from our target nets' list, which indicates that only two-level nets after asset need to be protected against probing attack. By further study, the target score threshold could be set to 0.125 to guarantee the two-level nets after assets are protected. So, we would recommend the designers to set the target score threshold to at most 0.125 to achieve the default protection against probing attacks or lower to involve more target nets if the budget allows.

Note that the asset should be identified by the chip designer as a user input in the anti-probing design flow as shown in Fig. 5.6. If one of the assets is not identified in the user input, the target net identification metric would not be able to recognize the nets that can leak information of the unidentified asset.

This chapter is mainly focusing on the protection against probing attack. So, the target net identification metric we proposed here is mainly developed to identify target nets in a probing attack. As a probing attacker, he/she tends to directly read out the critical information from the probed nets, which is the preferred nature of a probing attack without additional complicated analysis used in SAT attack or differential fault analysis (DFA). A universal target identification metric counting for a variety of attacks is out of the scope of this chapter because the principle to identify the target nets for various attacks differs a lot. Therefore, our metric is not intelligent enough to consider all types of attacks. For those nets that might be utilized to infer asset information through complicated mathematical analysis, e.g., the intermediate nets of an encryption/decryption process used in differential fault analysis (DFA) technique, they are not covered by the target net identification metric. However, these nets can also be protected in our methodology by declaring them in the user input as part of "asset" or setting the target score threshold to 0 to protect all fan-out nets of the keys against probing attack.

5.5.1.2   Shield Net Identification

One unique feature that distinguishes our proposed anti-probing physical design flow from previously proposed techniques is the adoption of internal functional nets of the design as shield to protect target nets against probing attack. Existing active shield countermeasures are vulnerable to bypass attacks and reroute attacks [14–16] because the shield at the top-most layer is relatively easy to access and manipulate. In addition, more advanced active shields require cryptographically secure pattern generators [3], which themselves are sources of vulnerability and additional overhead. In contrast, utilizing internal functional nets provides the following major advantages. First, they will be routed within internal layers of the chip and therefore far more difficult to bypass and reroute. Second, the design itself will generate these signals alleviating the need for pattern generation, which will save the major area overhead introduced by active shield pattern generation. In this design, we develop a technique for identifying which internal nets can be utilized as shielding nets (covering nets). We define the following five requirements along with associated metrics as follows:

- *Target score*: The shield nets should not carry any asset information since they are not protected and could be probed. The prior target score approach can be inverted to identify nets that carry the least sensitive information.
- *Toggle frequency*: The shield nets should have a relatively high toggling rate so that an attacker cannot replace them with a constant value after cutting them.
- *Switching probability*: It should be difficult to predict the signals on shield nets, which requires the probability of the net being 1 or 0 to be balanced.
- *Controllability*: The attacker should not have control over the shield nets. Otherwise, the shield can be replicated with the controlled value, allowing the attacker to freely perform the attack. The SCOAP controllability value [4] can be used for this feature and should be as high as possible.
- *Delay slack*: Using internal nets as shield nets should not impact the critical path delay and the design's performance. As discussed in later sections, shield nets will need to be extended and moved to cover target nets in the design, which will hurt the timing of the paths that the shield net belongs to. Hence, they should not lie on critical paths.

For each of the aforementioned shield requirements, a threshold value of corresponding metric should be determined to maximize the coverage on target nets and minimize the vulnerabilities and impacts from shield nets. The final shield candidate nets will be the intersection of the five net collections which satisfy the threshold values for each shield requirement.

5.5.1.3 Best Shield Layer

After appropriate shield nets are identified, the metal layer in the chip layout to route these shielding nets needs to be determined. In this study, we build two types of shield structures: single-layer shield and two-layer parallel shield. For single-layer shield, we assume a milling scenario using FIB technology as shown in Fig. 5.9, where colored bars are used to represent cross sections of metal wires on different routing layers. For the sake of argument, assume target wires (red) in the figure

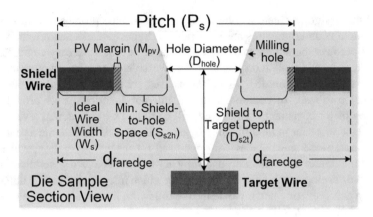

Fig. 5.9 Calculations for shield security and $d_{faredge}$

are on layer n and shield wires (purple) on layer $n + q$ and the attacker wishes to probe at one of the wires on target layer n to extract sensitive information. The hollowed-out cone (white) shown in the figure represents a cavity milled with FIB equipment. One known exploit on active shields is to create a reroute between identified equipotential points by circuit editing with FIB, so that the net would not become open when parts of the wires are removed [14]. This forces active shield designs to only use parallel wires with minimum spacing and widths [3] to maximize the attack complexity, because the shield with elbows (snake-like) may create a short path for reroute with a long section of the shield wire disabled.

From a layout point of view, active or analog shield designers are interested in the scenario where the attacker would make a mistake and leave a detectable footprint. To minimize the effect of the milling process, the attacker is likely to place the milling cavity in the center of two adjacent shield wires as shown in Fig. 5.9. To avoid affecting the normal signal transmission of shield wires, the attacker will avoid completely or partially cutting any shield wires. Further, a minimum space, S_{s2h}, is left between the shield wire and the milling cavity, as shown in Fig. 5.9, to minimize the effect of changed parasitic capacitance during the attack on the timing of shield wires. In order to account for the limitations of lithography and metalization as well, S_{s2h} is set to the same value with the minimum distance between metal wires as provided by the design rule of the technology. In addition, because of the process variation, the shield wires may be wider or thinner than the ideal wire width. Hence, to guarantee the minimum space between the shield wire and the milling cavity, an additional process variation margin (M_{pv}: typically 10% of the wire width) is added to the width of shield wire as shown in Fig. 5.9.

These restrictions create a maximal milling hole diameter limit on shield layer:

$$D_{\text{hole}} < P_s - W_s - 2M_{pv} - 2S_{s2h} \tag{5.3}$$

where P_s is the pitch size of shield layer, W_s is the ideal width of shield wires, M_{pv} is the process variation margin of shield wires, and S_{margin} is the minimal space between the shield wire and the milling cavity which can be determined by the minimal space between metal wires defined by the technology design rule . The milling hole diameter is determined by:

$$D_{\text{hole}} = \frac{D_{s2t}}{R_{\text{FIB}}} \tag{5.4}$$

where D_{s2t} is the depth from shield layer to target layer and R_{FIB} is the aspect ratio of FIB, which is defined as the ratio between FIB depth D_{s2t} and diameter D_{hole} as shown in Fig. 5.9. Therefore, the maximum FIB aspect ratio that the shield could protect against, which is termed as shield security [16], can be modeled as:

$$R_{\text{FIB,max}} = \frac{D_{s2t}}{P_s - W_s - 2M_{pv} - 2S_{s2h}} \tag{5.5}$$

The higher the *shield security* ($R_{\text{FIB,max}}$) value is, the better the single-layer shield is. The shield security can vary depending on shield layer, target layer, width of shield wire, and other layout technology parameters. Therefore, different technology libraries might lead to different shield security and different best shielding layer through Equation (5.5). Table 5.5 shows the shield security calculated from SAED32nm library. As we can see, shield layer 6 has the best shield security for target nets on layers 3 and 4 and also good for target nets on layers 1 and 2. Though shield layer 4 is better than layer 6 for target nets on layers 1 and 2 in terms of shield security, it requires to route all target nets within only two layers (layers 1 and 2) to take this advantage, which may cause serious routing congestion. Hence, layer 6 is the overall optimal shield layer with excellent shield security and sufficient space left for routing of target nets for single-layer shield designs. Therefore, in our single-layer internal shield implementation, shield nets are routed on metal 6, and target nets are routed under metal 4 (metal 4 included). Compared to the conventional active shield approach whose shield wires are routed on the top-most layer (metal

Table 5.5 Shield security in SAED32nm library

Max R_{FIB}	Shield layer							
Target layer	9	8	7	6	5	4	3	2
8	**0.46**	N/A						
7	**0.86**	0.64	N/A					
6	1.26	1.28	0.64	N/A				
5	1.66	1.91	1.28	1.81	N/A			
4	2.06	2.55	1.91	3.61	1.81	N/A		
3	2.46	3.19	2.55	5.42	3.61	4.41	N/A	
2	2.86	3.83	3.19	7.23	5.42	8.82	4.41	N/A
1	3.26	4.47	3.83	9.04	7.23	13.24	8.82	INF

9), the shield security for the best case of active shield (target on metal 1, shield on metal 9) is only 3.26, which is still less secure than the worst case of internal shield on M6 (target on metal 4, shield on metal 6) whose shield security is 3.61. In addition, the internal shield routed on metal 6 is more resistant to reroute attack where a shield path is duplicated between two equipotential points and bypass attack where the shield is bypassed by leveraging the space between adjacent shield wires, since the wires beyond shield layer (layers 7, 8, and 9) become huge obstacles to the attack.

Shield security is a very simple and useful metric to determine the best layer for single-layer shield. However, it might not be appropriate for multilayer shield structures, e.g., two-layer parallel shield, because adding an extra layer of shield might not increase the maximum FIB aspect ratio that the shield can protect against resulting in the same shield security value. Though multilayer shield might improve the protected ratio against a specific FIB, as long as this ratio is not 100%, the shield security will not be improved because it requires full protection. Therefore, we propose the *shield coverage* metric to determine which layers are good for two-layer parallel shield.

Let's consider the two-layer staggered parallel shield on M6 and M8 as shown in Fig. 5.10. The pitch size on M8 is twice of M6 as defined in SAED32nm library, and they have 50% offset to maximize the protection. The *shield coverage* is defined as:

$$\text{Coverage} = \frac{\text{Protected Region}}{\text{Period}} = \frac{\text{Period} - \text{Exposed}}{\text{Period}} \tag{5.6}$$

The *Period* is the pitch size of the upper shield layer (P_{m8}) because typically the upper layer has a larger pitch size than the lower layer. The *Exposed* is the region on target wires that is free to probe without triggering the shield alarm, which can

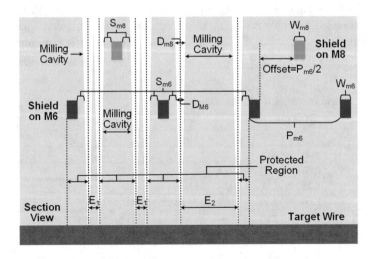

Fig. 5.10 Section view of a two-layer staggered parallel shield

be calculated as:

$$\text{Exposed} = 2 \times E_1 + \left(\frac{P_{\text{upper}}}{P_{\text{lower}}} - 1\right) \times E_2 \tag{5.7}$$

where P_{upper} and P_{lower} are the pitch size of upper shield layer and lower shield layer (P_{m8} and P_{m6} in Fig. 5.10, E_1 and E_2 are two types of exposed region as shown in Fig. 5.10) and are defined as:

$$E_1 = \frac{1}{2}P_{\text{lower}} - \frac{1}{2}(W_{\text{upper}} + W_{\text{lower}}) - (S_{\text{lower}} + S_{\text{upper}}) - \frac{1}{2}(D_{\text{upper}} + D_{\text{lower}}) \tag{5.8a}$$

$$E_2 = P_{\text{lower}} - W_{\text{lower}} - 2S_{\text{lower}} - D_{\text{lower}} \tag{5.8b}$$

where W_{upper} and W_{lower} are the metal width of upper shield layer and lower shield layer (W_{m8} and W_{m6}), S_{upper} and S_{lower} are the space between shield wire and milling cavity (S_{m8} and S_{m6} which can be determined by the minimal metal space defined by the technology design rules), and D_{upper} and D_{lower} are the milling cavity diameter on the upper shield layer and lower shield layer (D_{m8} and D_{m6}) which can be calculated using Equation (5.4).

The higher the *shield coverage*, the better the two-layer parallel shield design. As illustrated in Equations (5.7) and (5.8), the *shield coverage* depends on many factors defined by the technology and the selection of shield and target layers. Figure 5.11 shows the shield coverage of different two-layer staggered parallel shield designs using SAED32nm library. From the figure, we can see that all two-layer shield designs perform better than the single-layer shield on M6 design (black curve) especially when R_{FIB} is high. Though the two-layer shield on M5 and M6 is theoretically optimal for shield coverage, it brings a practical issue, routing

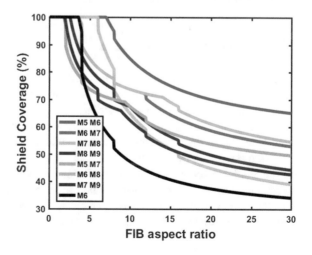

Fig. 5.11 Shield coverage of different shield designs in SAED32nm library

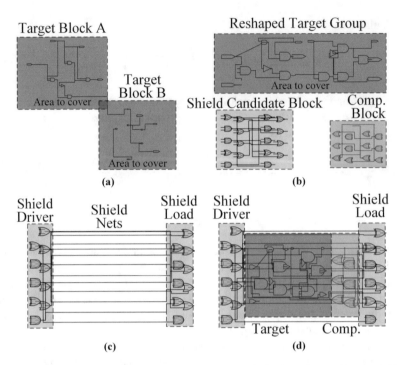

Fig. 5.12 (**a**) Irregular blocks (red) of sensitive target nets. (**b**) Reshape the sensitive target blocks to one regular rectangle block (red), shield candidate block (blue), and comparator block (green). (**c**) Shield gates (blue) are divided into shield nets' driver block and shield nets' load block. (**d**) Shield gates are placed surrounding the target and comparator blocks which will be covered by shield nets

congestion, due to the small pitch size on M5. Therefore, the shields on M6 and M8 (second best in shield coverage) are implemented for evaluation in later sections.

5.5.1.4   Floor-Planning Constraints

In conventional design flows, CAD tools perform floor-planning to optimize timing, power, and area. In an original design as shown in Fig. 5.12a, target nets and the blocks containing them (red) are distributed randomly throughout the design. It is neither easy nor efficient to protect them with such placement. It might also require more shield nets than available. A more advantageous approach is to constrain them into a regularly shaped region, e.g., a rectangle, as shown in Fig. 5.12b. This can be implemented by enumerating all gates connected to target nets and then creating a floorplan group to constrain their relative placements. The location of this floorplan group is chosen to remain as close to its original placement to reduce the impact on performance. The optimal dimensions of this floorplan group are found

by extracting all gates and nets involved into a sub-layout where only these gates
and nets are placed and routed.

The comparator is used to detect the attack by comparing the shield signal from
upper layer and another copy from lower layer. So the comparator nets should also
be protected like target nets because if these nets are tampered to maintain a static
value, the testability of the shield nets will be compromised. Hence, the comparator
gates (green) are constrained in a floorplan group besides the target block as shown
in Fig. 5.12d. The comparator is XNOR based, which is designed to output 1 when
there are no attacks, and optimized by design compiler for less area.

Unlike target nets, we divide the gates connected to shield nets into two
separate floorplan groups, *shield nets' driver group* and *shield nets' load group*,
as shown in Fig. 5.12c. Our proposed shield net identification metric ensures that
the performance overhead due to our constrained floor-planning is minimal. Both
shield nets' driver group and load group (blue) are constrained at opposite ends of
the expected shielding area (target and comparator block) as shown in Fig. 5.12d,
so that routing of shield nets crosses the target area and therefore provides vertical
protection from milling/probing. The shield nets' load group should be placed at the
comparator's side so that the received signals from shield nets could be compared
in the comparator. The driver and load gates of the shield nets are buffers. Larger
buffers are used for drivers to provide better driving capability for the long shield
nets.

5.5.1.5   Routing Constraints

In addition to creating floor-planning constraints, wire-routing constraints are also
necessary to protect the device against probing attacks with large aspect ratio FIB.
Aspect ratio of FIB is defined as the ratio between depth D and diameter d, as shown
in Fig. 5.13, of a milled cavity and is an important measure of FIB performance
[9]. A larger aspect ratio results in a milling cavity of smaller diameter on the
top-most exposed layers and therefore has less impact on the protective circuitry.
Section 5.5.1.3 has revealed the best shield layers for single-layer shield design
and two-layer parallel shield design to maximize the protection against probing

Fig. 5.13 Routing layer
constraints for target and
shield nets

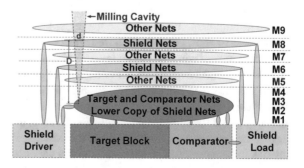

attack. Further, routing target nets in lower layer can also increase the chance to be protected from other non-shield internal function nets in the design. In this study, we route shield nets on M6 (M9 is the top layer) for single-layer shield design, route shield nets on M6 and M8 for two-layer parallel shield design, and route target nets and comparator nets under M4 (M4 included) to get an optimal protection as shown in Fig. 5.13. Further, another copy of the shield nets, which is the other branch of the driving buffers of the shield nets, is also routed under M4 to be compared with the genuine shield nets on the upper layers at the comparator.

5.5.1.6 Exposed Area (EA) Calculation

To assess the design's vulnerability to probing attacks, [9] proposed an exposed area metric by assuming that a complete cut of one shield wire is required for the detection of the attack. However, it is too conservative in several aspects. The first is assuming only complete cuts will be sufficient for detection. In reality, as soon as a minimum cross section of a cut wire to ensure correct signal transmission on the shield wires is violated, the attack is likely detected by active shield. Further, even if the milling cavity doesn't touch the shield wires, the changed parasitic capacitance due to the close distance between the shield wires and the milling cavity could possibly result in significant delays on shield wires especially in the low-power designs and thus trigger the alarm of an active shield by violating the setup time of flip-flops. In addition, the attackers will probably not try to challenge the sensitivity of the shield. They will always set a margin for the potential timing violation and also the FIB errors or operation mistakes. So, a space margin between the shield and the milling hole is necessary for attackers to improve the attack success rate. A more realistic model for detection is shown in Fig. 5.9, i.e., the probing attack can be detected if the center of milling exists within d_{faredge} from the far edge of the shield wire, where

$$d_{\text{faredge}} = \frac{D_{s2t}}{2R_{\text{FIB}}} + W_s + S_{\text{s2h}} + M_{pv} \qquad (5.9)$$

where D_{s2t}, R_{FIB}, W_s, S_{margin}, and M_{pv} are defined in Sect. 5.5.1.3. Equation (5.9) shows the possibility to find the area which milling center should not fall inside. We term this area the milling exclusion area (MEA). The desired exposed area (EA) will be its complement projected on the target layer.

Figure 5.14 shows how the exposed area (EA) can be found for any given target wire and covering wires on higher layers which are capable of projecting the milling exclusion area. Assuming the white region is the targeted wire at lower layer of a layout and the green and purple regions are the covering wires at upper layers above the targeted wire, the shaded region is the milling exclusion area (MEA), which indicates that if the milling center falls in this area, then the attack will be detected. Hence, the complement area of MEA is the desired exposed area that will not cause any risk to be detected. The exposed area can vary according to the different aspect

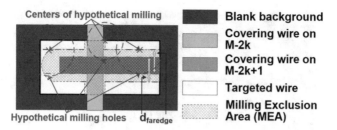

Fig. 5.14 Exposed area (EA) calculation

ratio of FIB, since the diameter of the cavities milled by FIB with different aspect ratio is different. Larger exposed area in the design represents more vulnerable to probing attacks.

5.5.2 Evaluation

In this section, the proposed FIB-aware anti-probing physical design flow is evaluated to find out how efficient the design flow can be and how much area in the design is vulnerable to probing attacks. For this purpose, layout of Advanced Encryption Standard (AES) and Data Encryption Standard (DES) crypto-cores are selected for the evaluation of the proposed design flow.

5.5.2.1 Implementation of Design Flow

The DES and AES modules used are from OpenCores [7]. They are described in register-transfer level (RTL) code and synthesized using Synopsys *Design Compiler* with Synopsys SAED 32 nm technology library. The layout of AES and DES modules are generated and constrained using Synopsys *IC Compiler*. The asset in the AES and DES modules is taken to be the encryption key (128 bits for AES and 56 bits for DES), which is hardcoded in the design.

The *target score* metric illustrated in Sect. 5.5.1.1 is used to identify the probing target nets in the AES and DES modules. The *target score* threshold value is set to 0.125 (target score for asset net is 1 and for non-asset net is 0), which results in nets within two levels after the asset nets are identified as probing target nets. Hence, 384 nets for AES and 200 nets for DES including key nets are probing target nets in the 2 designs. Further, gates connected to target nets are grouped and reshaped into a rectangular target block as shown in Fig. 5.15 (red). In addition, a 64-bit comparator is inserted in the AES and DES designs. Comparator gates are also grouped and reshaped into a rectangular block besides target gates block as shown in Fig. 5.15 (green).

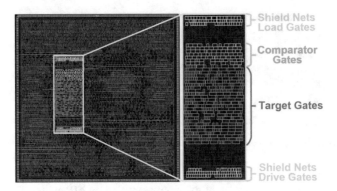

Fig. 5.15 Grouped and reshaped target gates, comparator gates, and shield gates in AES

Table 5.6 Threshold values for shield net identification in AES

Metric	Min.	Max.	Best	Percentage	Threshold
Target score	0	1	0	80%	<0.001
Togg. rate	0	0.06	0.06	40%	>0.0187
Delay slack	0.01	1.60	0.01	40%	<1.23
CC0 (SCOAP)	0	2532	2532	40%	>395
CC1 (SCOAP)	0	2081	2081	40%	>332
Probability	0	1	0.5	40%	0.22~0.78

Table 5.6 shows the metrics and threshold values used to identify shield nets in AES module to cover target block. The *Min.* and *Max.* columns show the minimum and maximum values measured in the design for each metric. The *Best* column indicates the optimal value as shield net for each metric. The optimal values for the metric of shield nets are the minimum values of target score and delay slack and maximum values of togg. rate, CC0, and CC1. The *Percentage* column presents percentage of all nets that are picked for each metric. The *Threshold* column indicates the threshold values for each metric, which are determined to offer a balanced trade-off between security and overhead. Hence, 136 nets in AES module and 118 nets in DES module, which meet all requirements of shield metrics, are identified as shield candidate nets for both designs. The final number of shield nets used for building the internal shield depends on the area on chip that needs to be protected against probing attack and the structure of the shield (single-layer or two-layer). In our implementation, 64 and 56 shield nets are used to build the single-layer internal shield for AES and DES, respectively. Therefore, in AES module, 64 driver gates and 64 load gates connected to the shield nets are reshaped into 2 groups, respectively, and placed at the opposite ends of target and comparator block as shown in Fig. 5.15 (yellow). Figure 5.16 shows the routing of target nets (a), shield nets (b), and their layer distribution (c)(d) in the AES layout. Target nets', comparator nets', and shield nets' copies are constrained in the reshaped target and

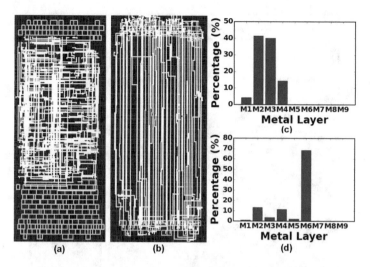

Fig. 5.16 (**a**) AES shield gates (yellow), target gates (red), and highlighted target nets under M4 (white). (**b**) AES shield gates (yellow) and highlighted shield nets on M6 (white). (**c**) Target nets' layer distribution. (**d**) Shield nets' layer distribution

comparator block and routed under M4 as discussed in Sect. 5.5.1.5. Most shield nets are routed on M6 to provide optimal coverage.

In addition to the single-layer internal shield designs, two-layer staggered parallel shield [10], which utilizes two routing layers to build the parallel shield with some offset between different layers, can provide better protection. Figure 5.17a shows an example of the two-layer staggered shield on M6 and M8. The pitch size on M8 is two times of the pitch size on M6 in SAED32nm library, which results in that the shield density on M8 being half of the shield density on M6. Fifty percentage offset is set between the shield wires on M6 and M8 to maximize the protection. Figure 5.17b shows the placement of target gates (red), comparator gates (green), and shield gates (yellow). Figure 5.17c, d, e shows the routing of target nets, shield nets on M6, and shield nets on M8, respectively.

Besides the baseline single-layer shield design and two-layer parallel shield design as illustrated above, we also implement four extra different designs for AES and DES, respectively, to show the high efficiency of our anti-probing physical design flow. Table 5.7 shows the description of implemented six different designs for AES and DES. Design No.1 is the original design using conventional place and route flow without any protection against probing attack. Design No.2 is the baseline single-layer shield on M6 as illustrated before. Design No.3 decreases the target score threshold from 0.125 to 0.01, which involves more target nets protected under the internal shield. Design No.4 includes those common fault injection target nets in the asset declaration, so that the nets vulnerable to fault injection attack are also protected under the shield. Design No.5 is the two-layer staggered parallel shield as shown in Fig. 5.17. Design No.6 is the conventional active shield design

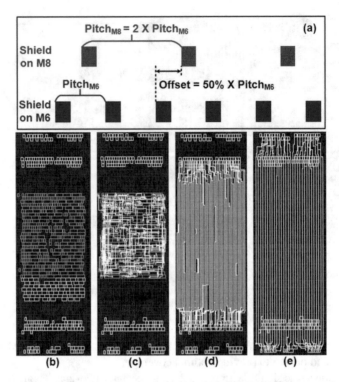

Fig. 5.17 (**a**) Diagram of two-layer staggered shield on M6 and M8. (**b**) Placement of target gates (red), comparator gates (green), and shield gates (yellow). (**c**) Routing of target nets. (**d**) Routing of shield nets on M6. (**e**) Routing of shield nets on M8

Table 5.7 Description of implemented designs for AES and DES

No.	Design	Notes
1	Original design	Conventional physical design flow
2	Single-layer shield I	Single-layer shield on M6
3	Single-layer shield II	Decrease target score threshold to 0.01 to include more target nets
4	Single-layer shield III	Include nets of fault injection positions in the asset
5	Two-layer parallel shield	Two-layer shield on M6 and M8
6	Active shield	Conventional active shield design

with a lightweight Simon cipher inserted as the shield signal pattern generator [3]. Figure 5.18 shows the diagram of a conventional active shield (**a**) and the layout of the implemented active shield on AES (**b**). The numbers of total gates, target nets, and target gates for different designs are listed in the $third$, $fourth$, and $fifth$ columns of Table 5.8.

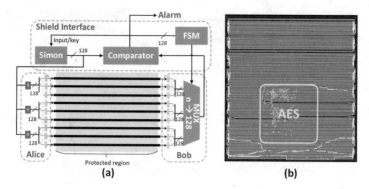

Fig. 5.18 (a) Diagram of a conventional active shield. (b) Layout of the implemented conventional active shield on AES

Table 5.8 Configuration and overhead of different AES and DES designs

Module	Design	Total gates	Target nets	Target gates	Timing	Power	Area	Routing
AES	2	10,293	384	512	0.32%	2.79%	0.74%	11.60%
AES	3	10,293	480	672	0.66%	3.66%	3.02%	14.80%
AES	4	10,293	535	833	0.66%	6.03%	3.17%	22.99%
AES	5	10,293	384	512	0.34%	4.90%	1.44%	17.77%
AES	6	28,048	384	512	3.95%	439.83%	402.31%	407.40%
DES	2	8882	200	432	1.18%	0.75%	0.51%	10.39%
DES	3	8882	376	637	4.55%	1.38%	0.50%	13.41%
DES	4	8882	481	783	4.55%	1.67%	0.80%	21.16%
DES	5	8882	200	432	1.18%	2.83%	1.78%	20.85%
DES	6	26,637	200	432	3.64%	365.17%	413.91%	556.54%

Table 5.8 also shows the timing, power, area, and routing overhead of all these designs compared to the original AES and DES without any constraints. Note that design No.2 and No.5 have the same number of total gates, target nets, and target gates, but they are implemented with different shield structure as shown in Table 5.7. As a result, their overhead is different. As we can see from the table, the overhead of the baseline single-layer shield (design No.2) is less than 3% for both AES and DES in timing, power, and area. In addition, the timing, power, and area overhead of all internal shield approaches (design No.2–5) are all less than 6% even with lower target threshold (design No.3) or including fault injection target nets (design No.4), which indicates that even if we increase our security standard to protect more sensitive nets against probing attack, the overhead is still acceptable and not increased too much. Further, if considering the overhead to an SoC, this amount of overhead could be completely ignored since AES or DES module is just a very small portion in an SoC. However, the conventional active shield approaches (design No.6) have ∼400% overhead in power, area, and routing, which is much larger than

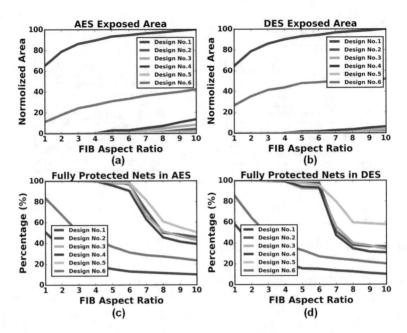

Fig. 5.19 (**a**) Exposed area in AES. (**b**) Exposed area in DES. (**c**) Percentage of fully protected target nets in AES. (**d**) Percentage of fully protected target nets in DES

our internal shield designs, because it requires the insertion of shield signal pattern generator and other supporting circuitry, e.g., FSM.

5.5.2.2 Exposed Area (EA)

The proposed internal shielding approach against probing attack is evaluated by the exposed area metric illustrated in Sect. 5.5.1.6. Figure 5.19a, b shows the normalized exposed area of all types of designs in Table 5.7 for AES (Fig. 5.19a) and DES (Fig. 5.19b). The exposed area is calculated across FIB aspect ratio from 1 to 10. As the FIB aspect ratio increases, the exposed area for all designs will also increase since $d_{faredge}$ decreases with larger FIB aspect ratio as shown in Equation (5.9), which results in smaller milling exclusion area (MEA) and thus larger exposed area (EA). By using the proposed anti-probing design flow, the exposed area of all internal shield designs (design No.2–5) can be reduced to 0 for both AES and DES when the FIB aspect ratio is low. Even with the advanced FIB (aspect ratio is 10), the exposed area of baseline single-layer shield (design No.2) and two-layer shield (design No.5) can be reduced at least to 5% and 2%, respectively, for both AES and DES. Figure 5.19c, d shows the percentage of fully protected target nets for all designs. We define a net that is fully protected as one that doesn't have any exposed area. From Fig. 5.19c, d, almost 100% of target nets for all internal shield designs (design No.2–5) are fully protected when $R_{FIB} \leq 6$, while less than 20% of target nets are fully protected for original AES and DES designs when $R_{FIB} = 6$. With the advanced FIB, there are still 50% and 60% of target nets fully protected under two-layer staggered shield (design No.5) for AES and DES, respectively, which is about five times more than the original AES and DES designs. For design No.2–4 with the same single-layer shield protection but different target nets' configuration and increasing overhead as shown in Table 5.8, Fig. 5.19c, d shows that they have similar security performance which indicates that our internal shield design flow can provide guaranteed protection with different target nets' configuration. Compared with internal shield designs, conventional active shield designs (design No.6) can only reduce the exposed area to ~40% and increase the number of fully protected nets by about two times, which is not efficient as shown in Fig. 5.19a–d.

5.6 Conclusions

Due to the proliferation of IC diagnosis, debug, and failure analysis equipment, the technological requirements to perform physical attacks on security critical ICs are dramatically declining. Further, considering the powerful capability of FIB-equipped adversaries, probing attacks have become an enormous threat to ICs for security-critical applications. In this chapter, we have reviewed the state-of-the-art and essential stages of probing attacks, as well as existing countermeasure techniques and their limitations. A CAD-based approach to protect the design against frontside probing threat is also illustrated. We expect this chapter to serve as a foundation for motivation and development of future methodologies that protect against probing and possibly other invasive physical attacks.

References

1. ARMInc: Building a secure system using TrustZone Technolog (2017). Available at http://infocenter.arm.com/help/topic/com.arm.doc.prd29-genc-009492c/PRD29-GENC-009492C_trustzone_security_whitepaper.pdf
2. Boit, C., Helfmeier, C., Kerst, U.: Security risks posed by modern IC debug and diagnosis tools. In: 2013 Workshop on Fault Diagnosis and Tolerance in Cryptography, pp. 3–11 (2013). https://doi.org/10.1109/FDTC.2013.13
3. Cioranesco, J.M., Danger, J.L., Graba, T., Guilley, S., Mathieu, Y., Naccache, D., Ngo, X.T.: Cryptographically secure shields. In: 2014 IEEE International Symposium on Hardware-Oriented Security and Trust (HOST), pp. 25–31 (2014). https://doi.org/10.1109/HST.2014.6855563
4. Goldstein, L.H., Thigpen, E.L.: Scoap: sandia controllability/observability analysis program. In: 17th Design Automation Conference, pp. 190–196 (1980)
5. Ishai, Y., Sahai, A., Wagner, D.: Private circuits: securing hardware against probing attacks. In: Boneh, D. (ed.) Advances in Cryptology – CRYPTO 2003, pp. 463–481. Springer, Berlin/Heidelberg (2003)

6. Manich, S., Wamser, M.S., Sigl, G.: Detection of probing attempts in secure ICs. In: 2012 IEEE International Symposium on Hardware-Oriented Security and Trust, pp. 134–139 (2012). https://doi.org/10.1109/HST.2012.6224333

7. OpenCores: AES and DES benchmarks. Available at https://opencores.org/projects/aes_core

8. Quadir, S.E., Chen, J., Forte, D., Asadizanjani, N., Shahbazmohamadi, S., Wang, L., Chandy, J., Tehranipoor, M.: A survey on chip to system reverse engineering. J. Emerg. Technol. Comput. Syst. **13**(1), 6:1–6:34 (2016). https://doi.org/10.1145/2755563

9. Shi, Q., Asadizanjani, N., Forte, D., Tehranipoor, M.M.: A layout-driven framework to assess vulnerability of ics to microprobing attacks. In: 2016 IEEE International Symposium on Hardware Oriented Security and Trust (HOST), pp. 155–160 (2016). https://doi.org/10.1109/HST.2016.7495575

10. Shi, Q., Wang, H., Asadizanjani, N., Tehranipoor, M.M., Forte, D.: A comprehensive analysis on vulnerability of active shields to tilted microprobing attacks. In: 2018 Asian Hardware Oriented Security and Trust Symposium (AsianHOST), pp. 98–103 (2018)

11. Skorobogatov, S.: Physical attacks on tamper resistance: progress and lessons. In: Proceedings of 2nd ARO Special Workshop Hardware Assurance, Washington, DC, 2011. Available at http://www.cl.cam.ac.uk/sps32/ARO_2011.pdf

12. Skorobogatov, S.: The bumpy road toward iPhone 5c NAND mirroring (2016). ArXiv preprint arXiv:1609.04327, 2016. Available at https://arxiv.org/ftp/arxiv/papers/1609/1609.04327.pdf

13. Tarnovsky, C.: Security failures in secure devices (2008). In: Proceedings of Black Hat DC Presentation, vol 74, Feb 2008. Available at http://www.blackhat.com/presentations/bh-dc-08/Tarnovsky/Presentation/bh-dc-08-tarnovsky.pdf

14. Tarnovsky, C.: Deconstructing a 'Secure' Processor (2010). Available at https://www.youtube.com/watch?v=w7PT0nrK2BE

15. Wang, H., Forte, D., Tehranipoor, M.M., Shi, Q.: Probing attacks on integrated circuits: challenges and research opportunities. IEEE Design Test **34**(5), 63–71 (2017). https://doi.org/10.1109/MDAT.2017.2729398

16. Wang, H., Shi, Q., Forte, D., Tehranipoor, M.M.: Probing assessment framework and evaluation of antiprobing solutions. IEEE Trans. Very Large Scale Integr. (VLSI) Syst. **27**(6), 1239–1252 (2019)

17. Wang, H., Shi, Q., Nahiyan, A., Forte, D., Tehranipoor, M.M.: A physical design flow against front-side probing attacks by internal shielding. IEEE Trans. Comput.-Aided Des. Integr. Circuits Syst. **39**(10), 1 (2019)

Chapter 6
Optical Inspection and Attacks

6.1 Introduction

In the past two decades, the desire for higher yield, faster failure analysis (FA), and fault localization have catalyzed tremendous progress in FA tools and physical inspection methods. Optical inspection techniques, such as photon emission (PE), laser voltage/electro-optical analysis, and laser fault injection, are examples of silicon debugging and diagnosis methods. Optical inspection methods are based on the well-known principle that silicon is transparent to near-infrared (NIR) photons performed from chip backside, i.e., the silicon substrate. However, an adversary can use the same approaches to localize and probe device assets and sensitive information, threatening the confidentiality and integrity of the device [2, 3, 40]. Aside from device assets such as encryption keys or firmware, an SoC may store the biometric, financial, and medical information of end user, company secrets, or intellectual property (IP). Therefore, SoC attack and inspection techniques have become an open issue for root of trust of the hardware. This is especially evident for optical attacks, which have been demonstrated as early as 2002 when a group of researchers flipped stored bits in a static random-access memory (SRAM) using a low-cost laser pointer and camera [31]. The list of optical attacks continued to grow since the demonstration of this attack.

The optical attacks are expected to experience another surge for several reasons. Metal layers and protective meshes designed to guard against probing attacks protect the frontside of dies. However, an adversary can exploit silicon transparency to near-infrared (NIR) photons to sidestep these security measures because silicon substrates remain unprotected and can be accessed by de-packaging or de-lidding. Besides, higher I/O pin count and compact design requirement for SoC compelled the semiconductor industry to shift toward flip-chip packaging. Unlike traditional dual in-line packaging (DIP), in flip-chip, the die is placed in an upside flipped orientation in the package, leaving the silicon substrate unprotected.

The objectives of this chapter are, first, to introduce readers to different classes of optical attacks. We will discuss physics as well as the data analysis approaches for each technique. Second, we explain the procedure involved in optical attacks. Attack models are also discussed. Third, we present an in-depth analysis of the security vulnerabilities caused by backside optical attacks and possible countermeasures.

6.2 Taxonomy of Optical Attacks

Optical Attack: An optical attack uses photons emitted by transistors, photons modulated by transistor activity, or perturbations in circuit activity due to photon stimulation to leak sensitive information. Due to continuously shrinking feature sizes and increasing device complexity, invasive attacks such as electrical probing have become more time-consuming and labor-intensive. For deep sub-micron processes, semi-invasive optical attacks can be launched with less expensive tools and lower time-cost than invasive attacks.

In this section, we present a taxonomy of optical attack techniques. Optical inspection/attack methods can be categorized into three major classes depending on stimulation and monitoring methods. They are (a) photon emission (PE), (b) electro-optical/laser voltage techniques, and (c) laser stimulation. Figure 6.1 depicts the taxonomy of different optical attack methodologies.

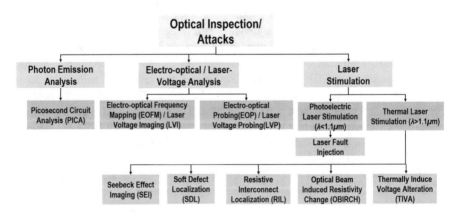

Fig. 6.1 Taxonomy of optical attacks/inspection techniques

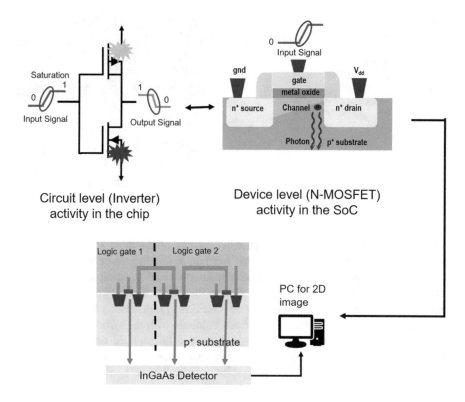

Fig. 6.2 Setup used for photon emission analysis. During logic gate switching, the transistors emit photons in saturation region. The emitted photons are captured using an InGaAs detector, and a 2D mapping of the device activity is generated

6.2.1   Photon Emission Analysis (PEA)

PEA is the simplest form of optical analysis (see Fig. 6.2). Initially, PEA was developed as a functional analysis and defect localization technique on ICs based on transistor switching activity. Since transistors emit photons without any external stimulation, PEA is a passive inspection method. During the switching of logic gates, the MOSFET transistors pass through the saturation for a brief period. The current carriers gain kinetic energy in the saturation region of the operation. This energy is released in the form of photons as hot-carrier luminescence. The hot-carrier luminescence occurs near the pinch-off region of the space-charge region in the MOSFET.

Photons are captured with a suitable detector, such as a Si-CCD, or InGaAs detector. Next, the signal from the detector is fed to a computer to create a 2D

Fig. 6.3 Setup used for
photon emission analysis.
During N-type MOSFET
switching, emitted photons
are captured using an CCD
detector, and a 2D mapping of
the device is generated [33]

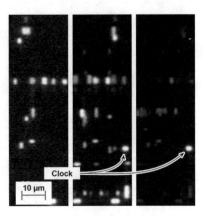

image to map the location of the switching activity of the logic gates. Figure 6.3
represents the photon emission collected from an FPGA implementation of arbiter
PUF [33]. The state of the logic gates can be interpreted depending on the presence
of emission at the logic gate location. Due to the higher mobility of electrons, the
intensity of emitted photons from n-type MOSFETs is significantly higher than that
of p-type MOSFETs. Therefore, emissions coincide with the switching activity of
n-type MOSFETs in logic gates. An adversary can threaten the confidentiality of
the on-chip assets by using the data dependency as mentioned above of hot-carrier
luminescence as a source of side-channel information [28, 37].

6.2.1.1 Photon Emission Detector

Three types of photon detectors are used for PEA: (1) charge-couple device (CCD),
(2) mercury cadmium telluride (MCT), and (3) indium gallium arsenide (InGaAs).
CCD detectors have been used extensively for photon emission microscopy, but
they have disadvantages. First, they must be cooled to $-50°C$ $-70°C$ to decrease
noise level due to dark current. They also can only be used to observe photons with a
wavelength of $300-1100$ nm because there is a sharp decrease in quantum efficiency
at the silicon bandgap (see Fig. 6.4). The operating voltage of the device defines
photon emission intensity. At low operation voltage, the wavelength of emission
photons lies in the infrared range. For photons with a wavelength of 900-1700 nm,
an InGaAs detector may be used with an emission microscope. MCT can detect
photons emitting between 800 and 2500 nm.

6.2.1.2 Picosecond Image Circuit Analysis (PICA)

PEA provides only the spatial switching activity of a transistor by accumulating
photons emitted by specific signals. Another approach, PICA [2, 28, 38], can be
integrated with PEA to provide temporal and spatial information about circuit

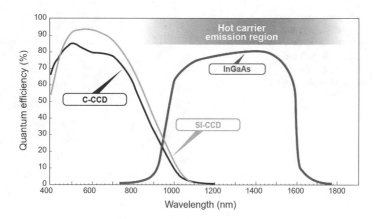

Fig. 6.4 Quantum efficiency of Si-CCD, InGaAs, and MCT detector [23]

switching activity and signal propagation. In PICA analysis, a circuit performs a function in a loop, and different parts of the circuit's operation are captured in each loop iteration. In the first iteration, the detector samples photons emitted during the first sampling window of the loop. In the next loop iteration, the sampling window is shifted, and this process is repeated. This technique allows the reconstruction of all switching activity for the looped function.

6.2.2 Electro-optical/Laser Voltage Techniques

Electro-optical methods are widely used for optical contactless probing and defect localization [15, 41]. Electro-optical techniques are an active approach for optically probing the transistor state through two well-known approaches: electro-optical probing (EOP) and electro-optical frequency mapping (EOFM) [17, 22, 36].

6.2.2.1 Source of EOP and EOFM Signal

EOP probes the activity of a single transistor or a logic gate. A laser stimulus is focused on a transistor. When the boundary between the silicon substrate and active regions of MOSFETs reflects the laser (see Fig. 6.5), the reflected beam will be amplitude- and phase-modulated by the electric field at the transistor drain and by the difference in densities of free carriers (e.g., depletion charge and inversion charge) at the transistor terminals. Because the transistor and logic gate states define electric field and free carriers' density, the modulation of the reflected beam can generate a time-domain waveform of transistor or logic gate activity [15].

Fig. 6.5 The laser beam gets
modulated by depletion and
space-charge region in the
transistor. The reflected laser
signal is fed to a spectrum
analyzer or oscilloscope for
optical contactless probing

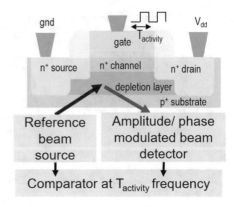

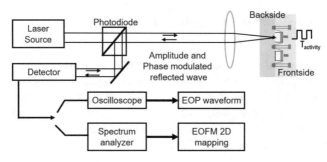

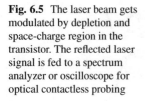

Fig. 6.6 Simplified illustration of contactless optical probing signal acquisition

An EOFM probing signal is generated using the same principle as EOP. The
difference between EOFM and EOP is that the former's focus is bound not to a
single transistor, but to any transistors in a region of interest (RoI) operating at a
frequency of interest. In EOFM, lasers scan the RoI in a raster fashion and collect
the modulated reflected beam. The beam is then analyzed in the frequency domain to
generate a 2D image representation of the activity mapping of the state of transistors
and logic gates.

6.2.2.2 EOFM and EOP Signal Measurement

Figure 6.6 shows the simplified contactless probing signal acquisition setup. A
measurement waveform is shown in Fig. 6.7. Two registers are connected to two
signals, sig_a and sig_b. The input value of sig_a and sig_b is bit "1" and bit "0,"
respectively. The circuit is connected to a reset signal which resets the registers
for an interval of T_{reset}. sig_a starts at the logic level low and then changes its
state, as soon as the time needed for the preceding calculation (T_{calc}) has elapsed.
As the interval for each consecutive power-on is constant, the interval length for
T_{calc} is equal to T_{reset}. Since sig_a switches at the frequency of reset signal, the

Fig. 6.7 Waveforms of the enable/clock (en), reset signal (rst), and two registers (sig_a and sig_b). The register sig_a receives a signal of bit "1," and the register sig_b receives a signal of bit "0"

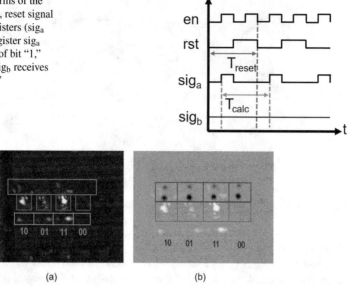

(a)   (b)

Fig. 6.8 (**a**) EOFM activity of D-flip-flops (blue rectangles) and buffers (green rectangles) operating at reset frequency (T_{reset} in Fig. 6.7). The value stored in each flip-flop is mentioned at the bottom of the buffer location. The stored value is identified based on the presence of activity (white dots) in the image. (**b**) Subtracted image of EOFM activity at reset and clock frequency. The black (red rectangles) and white dots (blue and green rectangles) represent the activity at clock and reset frequency, respectively [26]

registers (to be specific transistors) connected to sig_a also experience a change in the free carriers' density and modulate the laser which is exciting the register. The modulated signal is fed to a photo-diode, as shown in Fig. 6.6, which then converts the optical signal into an electrical signal. The electrical signal and the reset signal can be fed to an oscilloscope to generate a time-domain EOP waveform of register activity. Similarly, for EOFM analysis, the electrical signal from the RoI is fed to a spectrum analyzer where the signal is filtered with a narrow bandpass filter with center frequency at frequency of interest (e.g., the frequency of reset signal). Later, the filtered signal is fed to a computer to generate a 2D mapping of the activity of the registers. For example, in Fig. 6.8, EOFM activity of eight D-flip-flop, buffers, and clock, implemented in a PolarFire Microsemi FPGA, is shown. Figure 6.8a represents the EOFM activity mapping at the reset frequency of eight-bit registers and buffers. Figure 6.8b represents the subtracted image of EOFM activity mapping at clock and reset frequency [26]. Depending on the presence of white dots in the EOFM activity images, the value stored in the D-flip-flop or buffer can be determined.

Signals are often measured several times in a triggered loop to improve the signal-to-noise ratio (SNR) of electro-optical techniques. Frequency domain analysis may be used to reduce the signal acquisition time for EOFM compared to EOP.

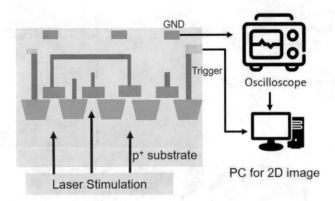

Fig. 6.9 Laser fault injection setup using photo-current stimulation and oscilloscope

Other optical probing approaches include laser voltage probing (LVP) and laser voltage imaging (LVI). LVP and LVI methods are equivalent to EOP and EOFM, respectively, except that the light source in laser-based techniques is incoherent.

6.2.3   Laser Stimulation (LS)

In an LS approach, an IR laser is applied from the silicon substrate to excite the circuitry of the device (see Fig. 6.9). Depending on the wavelength of the injected photons and silicon bandgap energy, two types of laser stimulation can be used: (a) photoelectric laser stimulation or (b) thermal laser stimulation.

6.2.3.1   Photoelectric Laser Stimulation (PLS)

The energy of photoelectric lasers can exceed the silicon bandgap (1.1 eV), enabling them to create electron-hole pairs. This effect is commonly referred to as photo-electric laser stimulation (PLS) [2, 31]. PLS can be used to flip the logic value of a CMOS gate by focusing the laser beam on the drain or source terminal of a transistor [31, 35]. This technique can inject faults into a circuit in what is known as a laser fault injection (LFI) attack. LFI attacks have been used for injecting faults in embedded devices, such as microcontrollers, FPGAs, and even physically unclonable functions (PUFs) [31, 35]. The success of laser fault injection depends on the wavelength, power, and exposure time of the incident photons [8].

Table 6.1 Classification, source of optical signal, and observable parameters of different optical attacks/inspection methodologies

Optical attack		Classification	Source of optical signal	Observable parameter
Photon emission		PEA, PICA	Energy emitted during switching of transistors	Spatial and temporal switching activity of logic gates
Electro-optical analysis		EOFM/LVI, EOP/LVP	Modulation of reflected laser caused by variation in free carrier density	1. Amplitude and phase modulation of reflected photons 2. Spatial and temporal switching of transistors
Laser stimulation	PLS	LFI	Induced carrier generation due to photoelectric effect	Change in the state of logic gate
	TLS	OBIRSH, TIVA	Seebeck effect, resistance change	Leakage current, short-open circuit

6.2.3.2   Thermal Laser Stimulation (TLS)

Thermal laser stimulation (TLS) is widely used in defect localization methods such as optical beam induced resistivity change (OBIRCH), soft defect localization (SDL), and thermally induced voltage alteration (TIVA). TLS uses a laser with a photon wavelength of 1.1 μm. This radiation induces phenomena such as the Seebeck effect and resistivity changes because its energy is lower than the silicon bandgap. Therefore, device parameters like voltage and current are affected [2, 18]. During TLS, a device is biased at supply voltage, and the current between supply pins is monitored with a current preamplifier. A computer samples the current preamplifier output, and a 2D map of the device's response is generated to localize current variations in the circuit.

The classification, source of optical signal, and observable parameters of different optical attacks and inspection methods are summarized in Table 6.1.

6.3   Threat Model and Potential Adversaries

A comprehensive threat model can be developed based on three potential security factors: (a) security assets, (b) attack approach, and (c) potential adversary.

6.3.1 Security Assets

The security assets of an SoC are the resources of value that are worth protecting from adversaries [1, 7]. A device's assets will vary depending on its application. We discuss a few typical examples of assets that are likely targets of probing attacks.

- **Device Key:** A few examples of device keys are encryption keys, eFuse configurations for silicon validation, and firmware authentication keys from original equipment manufacturers (OEMs). Compromise of these assets undermines a device's hardware root of trust.
- **On-Chip Protected Data:** Protected on-chip information, such as end-user information (e.g., login credentials or biometric data), protected firmware, and bitstreams, are appealing to malicious entities who seek to gain economic or competitive advantage. Therefore, leaking these sensitive data introduces integrity, confidentiality, and availability issues in a device.
- **Device Configuration, Digital Rights, and Checking Code:** Device configuration defines the information available to a particular user at different manufacturing stages of the device. Similarly, in an SoC, the digital rights management (DRM) policy defines access rights for different classes of users to protect valuable data from piracy [1, 21]. DRM can be implemented in either firmware or hardware. Hardware-based DRM provides better performance than software implementations in speed, power, memory usage, and security features. Leakage of device configuration or a breach in DRM violates the root of trust of the hardware.

6.3.2 Attack Approach

The ultimate goal of an optical attack is to acquire chip assets while simultaneously minimizing perturbations to a victim device. To achieve these goals through optical attacks, an attacker must complete the following steps (see Fig. 6.10):

1. **Acquire the Target Device:** Physical access to a device is necessary for an optical asset extraction. An adversary can acquire the chip from the open market or from an untrusted entity in the supply chain.
2. **Localize the Point of Interest (PoI):** An adversary can localize points of interest through partial or total reverse engineering. Therefore, the PoI localization approach is defined by the capability and information available to an attacker. For example, if the attacker is an untrusted foundry with access to the GDSII file, reverse engineering the netlist is sufficient for target localization. An end user with complete reverse engineering capabilities, e.g., a reverse engineering entity such as TechInsights (formerly Chipworks), would adopt an invasive reverse engineering approach to find the PoI. An attacker may also localize points of interest by identifying subsystems (e.g., memory, cryptomodule, communication

Fig. 6.10 Simplified illustration of different steps in optical attack approach

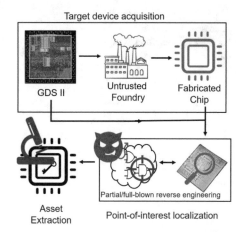

bus) using photon emission or imaging from the chip backside [9, 26, 37]. This partial/non-reverse engineering-based approach is more convincing for an attacker since it involves fewer FA tools and reduces both time and cost.

3. **Device-Under-Attack (DUA) Preparation and Optical Attack Approach Selection:** In optical attacks, an adversary monitors photon emission or modulation caused by transistor activity. Therefore, a clear path between the objective lens and transistors is necessary for optical attacks. In modern SoCs, the frontside of a die is covered by numerous metal interconnect layers that hide transistors from optical analysis. In some chips, active shield structures are implemented on the frontside of a chip to provide additional protection [24, 25, 40]. Chip backsides, on the other hand, are not covered by metal layers or protective mechanisms. For this reason, a chip's backside is an easy target for an adversary. Sample preparation for optical attacks entails exposing a chip's backside and optional substrate polishing. The greatest challenge to optical attacks is keeping the device under analysis (DUA) running.

Sample preparation of flip-chips is easy as accessing silicon substrate requires only lid removal and surface cleaning with a reagent like isopropyl alcohol (IPA). In contrast, sample preparation for regular DIP/BGA chips requires package removal by mechanical polishing or acid etching Fig. 6.11. Polishing or acid etching may damage chip bond wire, crack dies, or cause non-uniformity of the chip backside.

Chip backside thinning may also improve optical resolution [19]. Photons get absorbed while traveling through bulk silicon [10], so chip backsides must be polished. Mechanical backside polishing of die backsides must always be performed before LFI because of the low penetration depth of photoelectric photons. Anti-reflective coatings improve the quality of the polished surface for optical analysis. Silicon substrate is generally thinned down to 30–$10\,\mu$m depending on the optical analysis approach. Depending on the PoI, e.g., cache or register, an attacker then can choose the appropriate approach.

Fig. 6.11 Scanning electron microscope image of a backside thinned Intel SGX chip

Fig. 6.12 Circuit board placed under the lenses in a LSM [14]

4. **Necessary Equipment:** Depending on the attack approach, an adversary may require various FA tools. Recent advancements in FA instruments have made advanced microscopes capable of optical analysis more accessible. Laser scanning microscopes (LSM) may be used for electro-optical, TLS, and PLS analysis. A photon emission microscope (PEM) may be used for PEA. Various low-cost laser fault injection tools have been developed to facilitate fault injection attacks. An LSM and PEM have also been incorporated into a single FA tool (e.g., PHEMOS-1000 [23]) to provide a single solution for different optical debugging techniques (see Fig. 6.12).

In addition to equipment, access to backside polishing tools is necessary for a successful optical attack. Polishing can be completed in a non-selective or selective manner. For non-selective polishing, parallel lapping polishing tools or polishing grinders can be used. CNC machines and milling tools are typically used for selective policing.

6.3.3   Potential Adversaries

To protect their devices from optical attacks, SoC designers must be aware of potential adversaries, understand adversary capabilities, and enumerate security on their devices.

Adversaries can be classified into two major categories:

1. **Privileged Adversaries:** In a horizontal supply chain, untrusted entities (e.g., offshore foundries and third-party IP vendors) have access to chip design assets such as register-transfer level (RTL) IP, GDSII, and design-for-test architecture. Any entity with access to the design, verification, or fabrication process is considered a privileged adversary [24, 25]. SoC designers, foundries, and third-party design services [7] are all examples of privileged adversaries. This class of adversary may also acquire access to communication protocols, security policies, or debugging backdoors. Access to IP and GDSII files enables privileged adversaries to localize PoIs easily by reverse engineering netlists or exploiting DFT structures or firmware for asset extraction.

2. **Unprivileged Adversary:** End users only have access to fabricated chips and do not have control over design or/and fabrication and are therefore classified as unprivileged adversaries. Unprivileged adversaries are further classified based on their level of expertise:

   • **Expert Hardware Reverse Engineer:** The attacker is assumed capable of full-blown reverse engineering and has access to advanced FA tools. They may localize PoIs through netlist or firmware extraction or via reverse engineering.

   • **Hardware Intruder/Naive Reverse Engineer:** A hardware intruder leverages chip datasheets; expertise in firmware analysis; access to simple FA tools such as a polisher, LSM, or PEM; and social engineering to expose the PoI. They may employ partial reverse engineering as well. This class of attackers is the most dangerous due to their discreet nature.

Table 6.2 summarizes different classes of adversaries and their capabilities.

Table 6.2 A comprehensive taxonomy for different classes of the potential adversary in the optical attack. This taxonomy is also extendable for physical attacks too

Potential adversary		Example of adversary	Capability[a]	Available asset/knowledge
Privileged adversary		Untrusted foundry	1. Netlist reverse engineering tool 2. Failure analysis expertise and tool for full-blown reverse engineering	GDSII, design-for-test access
		SoC designer/rogue employee (rogue insider)	Netlist reverse engineering tool	RTL netlist, firmware, DFT access
		Third-party design service provider	Netlist reverse engineering tool	RTL netlist, GDSII, DFT access
Unprivileged adversary	Expert hardware reverse engineer	End user, untrusted distributor	1. FA expertise and tool for full-blown reverse engineering 2. Netlist and firmware reverse engineering capability	SoC, documentation of the chip
	Hardware intruder/naive reverse engineer	End user	1. Expertise about the firmware/software used in the device 2. Simple FA tools, e.g., optical microscope, mechanical polisher	

[a] = It is assumed that all adversaries have access to laser scanning microscope, photon emission microscope, and sample preparation tools

6.4 Security Threat of Optical Attacks

Optical attacks can impose threats to the integrity, confidentiality, and availability of various assets [10]. To understand the full range security threat of optical attacks, we elaborate on the vulnerability of different circuit components.

PEA is typically used to expose keys in cryptographic ciphers. The combination of spatial information from PEA and temporal information from PICA enables an adversary to extract PoIs to probe logic in microcontrollers, SoCs, and FPGAs [16, 28, 37]. Contactless optical probing, e.g., EOP and EOFM, has similarly been used to localize and probe bitstream encryption circuitry in FPGAs [36] and to probe logic gates (both sequential and combinatorial logic (see Fig. 6.8)) responsible for protecting device IPs [26]. LFI attacks have also successfully extracted keys from AES and RSA circuits [29].

LFI, PEA, and optical probing allow an attacker to probe memory read/write circuitry and non-volatile memories such as flash and EEPROM [29]. The contents of battery-backed RAM (BBRAM), a form of NVM widely used in FPGAs, are vulnerable to probing attacks [18]. Volatile memories such as SRAM are also susceptible to PEA, LFI, and TLS [18, 28, 32] attacks. In SoC architectures, SRAM serves as cache memory which can be considered a primary point of interest (PoI) for attackers and developers of security measures. Though physical unclonable functions (PUFs) are tamper-evident against invasive physical attacks, they are vulnerable to non- and semi-invasive attacks like PEA and LFI [33, 35, 37].

PEA and EOFM can also be used to locate security-sensitive circuitry. For example, an ARM core on an AI chip can be located using PEA (see Fig. 6.13) [9]. Note that reflected light from the backside of the chip makes it easier to differentiate between distinct modules on the chip (see Fig. 6.14).

Since such semi-invasive attacks extract assets through run-time monitoring, optical attacks are effective against standard security primitives. For example, the integrity of the boot procedure in modern processors is protected by "secure boot" policies with a hardware root of trust. The security assets in such a system, e.g.,

Fig. 6.13 The ARM core in a commercially available chip localized from PEA [9]

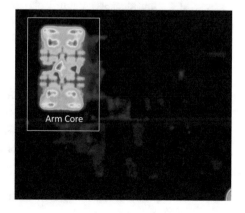

Fig. 6.14 The internal structure of a Microsemi FPGA collected with a 1.3 μm laser

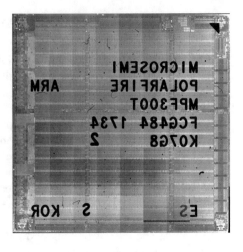

eFuse configurations or cryptographic keys, can be read from memory during boot using semi-invasive attacks. The vulnerability of embedded memories, caches, and registers to optical probing undermines security of the boot process. Sensitive information protected by "secured" embedded memory can no longer be considered "read-proof" against optical attacks.

6.5   Countermeasures Against Backside Optical Attacks

Two strategies may be employed to protect modern devices' assets: (a) prevention or (b) detection. Preventive strategies attempt to hinder asset extraction by shielding regions of interest (RoIs) from inspection or hiding circuit activity. Detection strategies monitor disturbances in a circuit that may result from optical intrusions, such as photons or temperature changes, and take action to protect systems that are deemed to be under attack.

Several countermeasures against backside optical attacks have been developed. These solutions can be classified into three categories: (i) material-based, (ii) device-based, and (iii) circuit-based countermeasures.

6.5.1   Material-Based Protection Mechanism

Material-based protection mechanisms block or detect infrared laser probing to safeguard the backside transistors. The practice of material-based protection of sensitive information within ICs is not new. Tamper-proof coating using silica resin, a hyper-oxidizing material, on the substrate has been proposed to protect ICs. The silica is oxidized when the circuit is analyzed during reverse engineering processes such as decapsulation, cross-sectioning, wet (acid/base) etching, or dry etching (plasma, reactive ion, or focused ion beam) [6]. The oxidized fillers release heat, which destroys the substrate and prevents further attacks. Tamper-resistant structures that damage underlying IC layers have also been proposed as a less disruptive approach for coating-based protection [5].

A read-proof protective coating consisting of TiO_2 and TiN particles in an aluminophosphate matrix [39] has also been proposed as a preventive measure. The protective coating makes the IC opaque to laser light, hardens the exterior structure to prevent wet etching attacks on the silicon die, and introduces randomness in the layers similar to PUFs. Such randomness can create a unique fingerprint, which can then be used by a cryptography module to generate a unique key, which is never stored on the chip itself. Their countermeasure is immune to electrical probing using a FIB, but not optical attacks.

Material-based security solutions involve a physical protective shield. Shielding is challenging to implement in compact systems (e.g., smart cards), in systems that require extra cooling (e.g., graphic cards and CPUs), and in next-generation packaging methods such as flip-chips. As the semiconductor technology progresses toward smaller feature sizes, low-power designs, and compact IC packages, countermeasures based on shielding are becoming obsolete.

6.5.2   Device-Based Protection Mechanism

Device-layer solutions use existing semiconductor process and device layers to protect the backside of ICs. No extra material or shield is required outside the IC, so circuit applications are preserved and the device remains accessible to failure analysis. A p-n junction of the diffusion region located at the backside of the chip has been proposed as physical tampering detection approach [42]. Forward-biased p-n junctions illuminate (light-emitting diode (LED) mode) the backside silicon, whereas reverse-biased p-n junctions can detect (photo-diode mode) any light reflected from the backside substrate. Adding a light-modulating structure at the silicon substrate can allow intrusion detection mechanisms to identify tampering or backside polishing. However, in situations where backside polishing is not required, such a layer would raise no alarm. A layer coated with ITO-AG-ITO has been used to introduce angle-dependent reflections to mitigate this shortcoming.

In [30], nanopyramid structures are implemented in selective areas inside the chip to prevent optical attacks by scattering the reflected laser beams (see Fig. 6.15).

A packaging method that includes a backside shield with deeply etched blind holes and a metal lining [4] can be added to the device. The etched holes weaken the chip, leading to damage if mechanical stress is introduced by polishing or a plasma FIB. The metal layer, which is opaque to IR, prevents laser fault injection attacks.

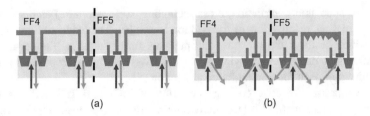

Fig. 6.15 Incident laser beam is scrambled by nanopyramids to disturb the LVP signal for optical probing [30]. (**a**) Incident and reflected laser. (**b**) Reflected scattered laser

6.5.3   Circuit-Based Protection Mechanism

Circuit-level countermeasures include logic gates that are resilient and tolerant to fault attacks, design flows that increase time and cost of optical invasion, and sensors that detect various steps in backside attacks.

Single-event upset (SEU)-tolerant flip-flops, known as Quatro latches, and Quadded logic (QL) techniques have been proposed [11, 13, 27] as a means of making circuits fault-tolerant. In Quadded logic, four concentric logic units are used along with a round-robin time-dependent rotating schedule to detect fault injection attempts [27]. A ring oscillator (RO)- and phase-locked loop (PLL)-based laser fault detector was proposed in [12]. The RO frequency changes as a result of laser stimulation. The PLL allows protection logic to determine whether the frequency change is large enough to indicate potential tampering. However, this approach poses multiple challenges. First, the selection of the threshold value is nontrivial. Second, though the PLL can accommodate device aging effects, the RO itself is a "power killer." Finally, since optical fault injection is a local attack, the expected spatial coverage required by this detector will cause significant area and power overhead. To detect laser stimulation, monitoring approaches based on physical unclonable functions have been proposed (see Fig. 6.16). This approach uses a combination of ring oscillator networks (RON) and RO sum PUFs to detect changes in frequency of ring oscillators by monitoring the state of counters of two adjacent ROs and RO sum PUFs.

Since chip backside polishing is used to improve the resolution of optical attacks and reduce photon absorption during laser fault injection, backside polishing detectors (BPD) can also detect attacks. BPDs monitor the parasitic capacitance of through-silicon vias (TSVs) to detect backside polishing. This technique assumes that a circuit is designed to operate in the MHz frequency range where other parasitic effects such as resistive and inductive effects are negligible. The capacitance of the TSVs decreases when the substrate is thinned (due to polishing attempts) and can be monitored with a delta meter to create a unique signature of the unpolished chip [20]. The major disadvantage of this countermeasure is that optical probing can be performed on many flip-chip ICs without any polishing. Also, the TSV process is very complicated and costly for low-cost, bulk manufacturing that is still fabricated in large transistor technology nodes.

Fig. 6.16 (a) Ring oscillator network to detect LVP attacks. (b) The modified architecture of the RO sum PUF network uses PUF output to detect optical attack [34]

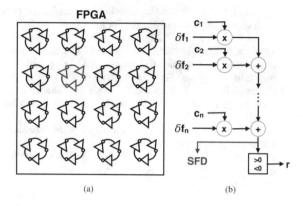

(a)   (b)

6.6 Conclusions

In this chapter, a comprehensive study of various optical attack techniques and security threats to SoCs was presented. Optical attacks are a threat to many existing security assets in SoCs. They exploit the lack of security of chip backsides to monitor circuits at run-time and compromise security assets. We have also classified potential adversaries based on their capability and their access to chip design assets. This classification of adversary facilitates a practical threat model for SoC assets by identifying attack surfaces available for optical debugging approaches. We have also discussed state-of-the-art countermeasures for optical attacks and the limitation of those protection schemes. There is no "silver bullet" against optical attacks, but understanding attack approaches and potential adversaries aids the development of new security schemes that protect chip backsides from unauthorized access.

References

1. Arm, A.: Security technology-building a secure system using trustzone technology. ARM Technical White Paper (2009)
2. Boit, C., Helfmeier, C., Kerst, U.: Security risks posed by modern ic debug and diagnosis tools. In: 2013 Workshop on Fault Diagnosis and Tolerance in Cryptography, pp. 3–11. IEEE (2013)
3. Boit, C., Tajik, S., Scholz, P., Amini, E., Beyreuther, A., Lohrke, H., Seifert, J.P.: From ic debug to hardware security risk: The power of backside access and optical interaction. In: 2016 IEEE 23rd International Symposium on the Physical and Failure Analysis of Integrated Circuits (IPFA), pp. 365–369. IEEE (2016)
4. Borel, S., Duperrex, L., Deschaseaux, E., Charbonnier, J., Cledière, J., Wacquez, R., Fournier, J., Souriau, J.C., Simon, G., Merle, A.: A novel structure for backside protection against physical attacks on secure chips or sip. In: 2018 IEEE 68th Electronic Components and Technology Conference (ECTC), pp. 515–520. IEEE (2018)
5. Byrne, R.C.: Tamper resistant integrated circuit structure (1994). US Patent 5,369,299
6. Camilletti, R.C., Haluska, L.A., Michael, K.W.: Tamper-proof electronic coatings (1995). US Patent 5,458,912

7. Contreras, G.K., Nahiyan, A., Bhunia, S., Forte, D., Tehranipoor, M.: Security vulnerability analysis of design-for-test exploits for asset protection in socs. In: 2017 22nd Asia and South Pacific Design Automation Conference (ASP-DAC), pp. 617–622. IEEE (2017)

8. Courbon, F., Loubet-Moundi, P., Fournier, J.J., Tria, A.: Increasing the efficiency of laser fault injections using fast gate level reverse engineering. In: 2014 IEEE International Symposium on Hardware-Oriented Security and Trust (HOST), pp. 60–63. IEEE (2014)

9. Dhwani, M., Tajik, S., Woodard, D., Asadi, N., Tehranipoor, M.: On the physical security of ai accelerators. In: International Conference on Physical Assurance and Inspection of Electronics (2019)

10. Green, M.A., Keevers, M.J.: Optical properties of intrinsic silicon at 300 k. Prog. Photovolt. Res. Appl. **3**(3), 189–192 (1995)

11. Han, J., Leung, E., Liu, L., Lombardi, F.: A fault-tolerant technique using quadded logic and quadded transistors. IEEE Transactions on Very Large Scale Integration (VLSI) Systems **23**(8), 1562–1566 (2014)

12. He, W., Breier, J., Bhasin, S., Miura, N., Nagata, M.: Ring oscillator under laser: potential of pll-based countermeasure against laser fault injection. In: 2016 Workshop on Fault Diagnosis and Tolerance in Cryptography (FDTC), pp. 102–113. IEEE (2016)

13. Jagannathan, S., Loveless, T., Bhuva, B., Wen, S.J., Wong, R., Sachdev, M., Rennie, D., Massengill, L.: Single-event tolerant flip-flop design in 40-nm bulk cmos technology. IEEE Trans. Nucl. Sci. **58**(6), 3033–3037 (2011)

14. Jain, A., Rahman, M.T., Guin, U.: Atpg-guided fault injection attacks on logic locking. In: Conference on IEEE International Conference on Physics Assurance and Inspection of Electronics (PAINE) (2020)

15. Kindereit, U.: Fundamentals and future applications of laser voltage probing. In: 2014 IEEE International Reliability Physics Symposium, pp. 3F–1. IEEE (2014)

16. Krämer, J., Nedospasov, D., Schlösser, A., Seifert, J.P.: Differential photonic emission analysis. In: International Workshop on Constructive Side-Channel Analysis and Secure Design, pp. 1–16. Springer (2013)

17. Liao, J.Y., Kasapi, S., Cory, B., Marks, H.L., Ng, Y.S.: Scan chain failure analysis using laser voltage imaging. Microelectron. Reliab. **50**(9–11), 1422–1426 (2010)

18. Lohrke, H., Tajik, S., Krachenfels, T., Boit, C., Seifert, J.P.: Key extraction using thermal laser stimulation. IACR Trans. Cryptogr. Hardw. Embed. Syst. **2018**, 573–595 (2018)

19. Lohrke, H., Zöllner, H., Scholz, P., Tajik, S., Boit, C., Seifert, J.P.: Visible light techniques in the finfet era: Challenges, threats and opportunities. In: 2017 IEEE 24th International Symposium on the Physical and Failure Analysis of Integrated Circuits (IPFA), pp. 1–6. IEEE (2017)

20. Manich Bou, S., Arumi Delgado, D., Rodríguez Montañés, R., Mujal Colell, J., Hernández García, D.: Backside polishing detector: A new protection against backside attacks. In: DCIS'15-XXX Conference on Design of Circuits and Integrated Systems (2015)

21. Nath, A.P.D., Bhunia, S., Ray, S.: Artifact: Architecture and cad flow for efficient formal verification of soc security policies. In: 2018 IEEE Computer Society Annual Symposium on VLSI (ISVLSI), pp. 411–416. IEEE (2018)

22. Ng, Y.S., Skvortsov, D.: Systems and method for laser voltage imaging state mapping (2014). US Patent 8,754,633

23. Photonics, H.: Emission Microscopy: Phemos-1000

24. Rahman, M.T., Asadizanjani, N.: Backside security assessment of modern socs. In: 2019 20th International Workshop on Microprocessor/SoC Test, Security and Verification (MTV), pp. 18–24. IEEE (2019)

25. Rahman, M.T., Rahman, M.S., Wang, H., Tajik, S., Khalil, W., Farahmandi, F., Forte, D., Asadizanjani, N., Tehranipoor, M.: Defense-in-depth: A recipe for logic locking to prevail. Integration (2020)

26. Rahman, M.T., Tajik, S., Rahman, M.S., Tehranipoor, M., Asadizanjani, N.: The key is left under the mat: On the inappropriate security assumption of logic locking schemes. In: Conference on IEEE International Symposium on Hardware Oriented Security and Trust (HOST) (2020)
27. Schiefer, P., McWilliam, R., Purvis, A.: Fault tolerant quadded logic cell structure with built-in adaptive time redundancy. Procedia CIRP **22**, 127–131 (2014)
28. Schlösser, A., Nedospasov, D., Krämer, J., Orlic, S., Seifert, J.P.: Simple photonic emission analysis of aes. In: International Workshop on Cryptographic Hardware and Embedded Systems, pp. 41–57. Springer (2012)
29. Schmidt, J.M., Hutter, M., Plos, T.: Optical fault attacks on aes: A threat in violet. In: 2009 Workshop on Fault Diagnosis and Tolerance in Cryptography (FDTC), pp. 13–22. IEEE (2009)
30. Shen, H., Asadizanjani, N., Tehranipoor, M., Forte, D.: Nanopyramid: An optical scrambler against backside probing attacks. In: ISTFA 2018: Proceedings from the 44th International Symposium for Testing and Failure Analysis, p. 280. ASM International (2018)
31. Skorobogatov, S.P., Anderson, R.J.: Optical fault induction attacks. In: International Workshop on Cryptographic Hardware and Embedded Systems, pp. 2–12. Springer (2002)
32. Stellari, F., Song, P., Villalobos, M., Sylvestri, J.: Revealing sram memory content using spontaneous photon emission. In: 2016 IEEE 34th VLSI Test Symposium (VTS), pp. 1–6. IEEE (2016)
33. Tajik, S., Dietz, E., Frohmann, S., Dittrich, H., Nedospasov, D., Helfmeier, C., Seifert, J.P., Boit, C., Hübers, H.W.: Photonic side-channel analysis of arbiter pufs. J. Cryptol. **30**(2), 550–571 (2017)
34. Tajik, S., Fietkau, J., Lohrke, H., Seifert, J.P., Boit, C.: Pufmon: Security monitoring of fpgas using physically unclonable functions. In: 2017 IEEE 23rd International Symposium on On-Line Testing and Robust System Design (IOLTS), pp. 186–191. IEEE (2017)
35. Tajik, S., Lohrke, H., Ganji, F., Seifert, J.P., Boit, C.: Laser fault attack on physically unclonable functions. In: 2015 Workshop on Fault Diagnosis and Tolerance in Cryptography (FDTC), pp. 85–96. IEEE (2015)
36. Tajik, S., Lohrke, H., Seifert, J.P., Boit, C.: On the power of optical contactless probing: Attacking bitstream encryption of fpgas. In: Proceedings of the 2017 ACM SIGSAC Conference on Computer and Communications Security, pp. 1661–1674. ACM (2017)
37. Tajik, S., Nedospasov, D., Helfmeier, C., Seifert, J.P., Boit, C.: Emission analysis of hardware implementations. In: 2014 17th Euromicro Conference on Digital System Design, pp. 528–534. IEEE (2014)
38. Tsang, J.C., Kash, J.A., Vallett, D.P.: Picosecond imaging circuit analysis. IBM J. Res. Dev. **44**(4), 583–603 (2000)
39. Tuyls, P., Schrijen, G.J., Škorić, B., Van Geloven, J., Verhaegh, N., Wolters, R.: Read-proof hardware from protective coatings. In: International Workshop on Cryptographic Hardware and Embedded Systems, pp. 369–383. Springer (2006)
40. Vashistha, N., Rahman, M.T., Paradis, O.P., Asadizanjani, N.: Is backside the new backdoor in modern socs? In: 2019 IEEE International Test Conference (ITC), pp. 1–10. IEEE (2019)
41. Yee, W.M., Paniccia, M., Eiles, T., Rao, V.: Laser voltage probe (lvp): A novel optical probing technology for flip-chip packaged microprocessors. In: Proceedings of the 1999 7th International Symposium on the Physical and Failure Analysis of Integrated Circuits (Cat. No. 99TH8394), pp. 15–20. IEEE (1999)
42. Zachariasse, F.: Semiconductor device with backside tamper protection (2012). US Patent 8,198,641

Chapter 7
Package Security

7.1 Introduction

The four primary functions of integrated circuit (IC) packaging are power distribution, signal distribution, heat dissipation, and package protection. These functions give IC packaging a crucial role in the reliability and assurance of microelectronic products. The early failure of chips can be caused by adversarial changes to packaging parameters by a malicious manufacturer. Furthermore, globalization of the IC packaging supply chain and the complexity of advanced semiconductor devices' packaging make modifications to packaging easy to perform and difficult to detect. Modern in-process inspection methods and accelerated cycling tests on finished chips are commonly used to monitor and prevent intentional or unintentional changes in packaging. However, it is still difficult for an end user to detect structural or packaging modifications before receiving a device. The time-consuming nature of modern reliability testing methods is not the only challenge in packaging reliability evaluation. The numerous packaging parameters involved in the entire process also need to be assessed. In this chapter, common IC packaging vulnerabilities and possible countermeasures are discussed.

7.1.1 IC Packaging Evolution

IC packaging was first introduced in the 1970s, near the beginning of the electronics era. Early packaging was developed for the sole purpose of encapsulating a die and establishing connections with a motherboard. While these functions still apply to modern chips, additional packaging features are required to meet the needs of continually advancing IC technology. For example, to accommodate high-performance and small-sized ICs, packaging techniques such as wafer-level packaging (WLP) were developed [17]. Similarly, system-on-chip (SoC) and system-in-packaging

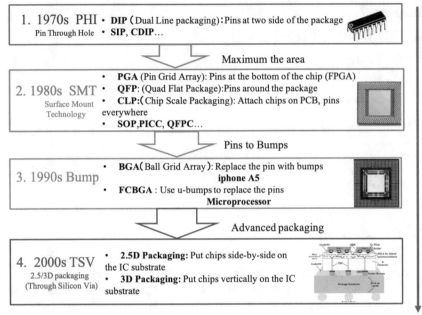

Fig. 7.1 IC packaging evolution (1970–now)

(SiP) technologies offer higher heat dissipation and lower power cost than traditional packaging approaches. The IC packaging evolution is depicted in Fig. 7.1.

The first type of IC packaging was dual in-line packaging (DIP), which placed pins at two sides of the packaging. However, the limited space to place pins in a DIP package meant these samples could only contain 4–64 pins. In the 1980s–1990s, the pin grid array (PGA) and ball grid array (BGA) packaging used surface mounting technology to place more pins at the bottom of the packaging [30]. Surface mounting technology allows the chip to directly attach to the motherboard, increasing the thermal transfer and reducing mechanical stress, meeting the requirements of high-performance chips.

SoC, the next improvement in the packaging approach, included different electronic circuits on a single die [16]. SoC was equally effective for both mixed analog and digital circuit packaging. In the same time period, flip-chip technology such as FCBGA was used to decrease the packaging's thickness further and enhance the thermal transformation [22]. While SoC packages had the benefit of consolidating several separate systems, they generally resulted in larger required die areas. This drastically decreased yield per batch and increased design costs. The multi-chip module (McM) packaging was proposed and developed in 1985 to overcome this challenge. However, the high heat generation and low yield of this technology limited its development. In 2000, new wafer fabrication and interconnection technologies such as silicon interposer and micro-bump appeared as emerging solutions for the abovementioned issues, leading to SiP as a more appropriate selection for high-performance chips. SiP includes homogeneous and heterogeneous chips which use an interposer or embedded bridges to communicate [16]. When a die is placed horizontally within an SiP package, it is known as 2.5D packaging, whereas a vertically positioned die results in 3D packaging.

In summary, modern IC packaging has a complicated structure and involves significant interplay between several distinct technologies. It not only uses traditional packaging technologies, such as DIP, PGA, etc., but includes advanced techniques such as SoC, 3D connections, and more in the process. Early manufacturing plants were not equipped with the tools required to leverage these advancements; this leads to outsourcing of multiple steps in the IC packaging process.

7.1.2 IC Packaging Supply Chain Vulnerabilities

The shift from traditional IC packaging techniques to advanced packaging methods required significant investments due to large-scale equipment and fabrication facility upgrades. Extensive research was necessary to develop advanced packaging methods and improve yield.

These improvements were prohibitively expensive, especially for smaller companies, many of which were forced to outsource their packaging process. Hence, the current market consists of two types of companies: (a) outsourced semiconductor assembly and test (OSAT) companies and (b) foundries.

OSAT companies are primarily responsible for IC packaging and testing. After they receive wafers from a foundry, OSAT companies test and mark "good" and "bad" dies on each wafer. For the wafer-level packaging, as shown in Fig. 7.2, the wafer will be mounted on a large-scale substrate to form the packaging before dicing. For other packaging types, good dies will be cut from the wafer and packaged. Automated test equipment (ATE) tests the electrical properties of finished chips to ensure that they did not suffer sample or interconnection damage during the packaging process. For IC packaging, the offshore fabrication trend is inevitable, especially for advanced IC packaging. As shown in Table 7.1, most OSATs are not

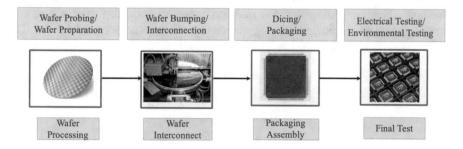

Fig. 7.2 Packaging supply chain of OSAT and foundry (wafer-level packaging)

Table 7.1 2018 & 3Q19 market share of top 8 OSATs

Company name	Country	2018 revenue (in M $)	2018 market share (%)	3Q19 revenue (in M $)	3Q19 market share (%)
ASE	Taiwan	5250	18.91	1321	22
Amkor	US	4316	15.62	1084	18.1
JCET Group	China	3604	13.14	1006	16.8
SPIL	Taiwan	2899	10.25	763	12.7
PTI	China	2257	7.89	566	9.4
Tongfu	China	1087	3.92	352	5.9
Tianshui Huatian	China	1071	3.85	324	5.4
UTAC	Singapore	788	2.81	183	3.1
Total	—	17,372	—	5599	—

located offshore. Amkor is the only American company in the top 8 OSATs, and its fabrication facilities are located offshore [1]. Recently, wafer foundries such as TSMC and Samsung have also entered the IC packaging market. Foundries offer their packaging service only for their in-house products.

Offshore fabrication is a significant source of vulnerability in the packaging process. These vulnerabilities are hard to avoid when the end users or IP owners lack control over the supply chain [18]. An estimated 10% of electronic components sold on the global market are not genuine [7]. As a result of increased outsourcing, quality assurance procedures that are not rigorous and widespread can cause unspecified damages to economics or critical infrastructure. Although existing reliability and verification methods are robust and well developed, advanced packaging still suffers from assurance and security problems.

7.2 IC Packaging Reliability Verification

Some standards and characterization methods are useful for verifying the authenticity and reliability of IC packaging, such as AS6171. However, existing characterization approaches lack in situ, real-time, and non-destructive reliability detection methods. Not all packaged chips undergo extensive reliability and assurance verification steps. Hence, identifying malicious modifications is nontrivial. An IC with an unreliable packaging may cause early failure of the device. For instance, in 2006, the failure of Sony CCD chips at high temperatures is suspected to be caused by improper package sealing and vaporization of bonding components [2]. Several detailed reliability verification methods will be introduced in this section.

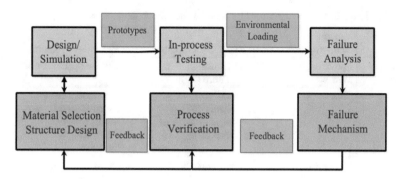

Fig. 7.3 IC packaging reliability verification flow

7.2.1 Reliability Verification Process

Current reliability testing methods do not depend on static analysis of package properties. Instead, samples are subjected to temperature and moisture stress experiments [6]. Physical inspection is performed on failed chips to gain insight into the cause of failure, and designs or material compositions are adjusted as necessary. This process is repeated until desirable performance is achieved. The verification process can be separated into three main stages, design, in-process testing, and failure analysis, as presented in Fig. 7.3 [29].

1. **Design:** During the design stage, several material property simulations are performed to determine which material compositions are best suited to the current sample. This simulation also ensures that whatever composition is selected does not interfere with chip functionality. After a package is designed, several prototypes are fabricated and tested using in-process methods.

2. **In-Process Testing:** In-process testing serves to test the integrity of packaging materials and the packaging process. Properties of test packages are compared to simulation results to ensure that simulations accurately reflect packaging characteristics. This comparison of packaging characteristics with models used in simulation verifies that the materials selected will not fail before stress testing. Such premature failure might occur if a vendor does not properly store their components, resulting in microfractures not caused by the intentional stress loading during later failure analysis [6]. In other words, this test ensures that defects found during failure analysis are solely from stress testing and not from the manufacturing or design process.

3. **Failure Analysis:** During failure analysis tests, several forms of mechanical stress will be applied to the chips to simulate real working conditions. Several factors are altered at this point, including temperature, moisture, and duration of time under strenuous conditions. Physical inspection methods will locate failures

Table 7.2 Reliability problems caused by material and fabrication process

	Material composition	IC Packaging Properties	Reliability Problems
Encapsulant	Epoxy Resins Matrix	Glass Transition Temperature (Tg)	High temperature reliability
	Fillers	Thermal Transport Coefficients (TC) Coefficient of Thermal Expansion (CTE)	Disconnection Delamination Voids, Crack Corrosion Warpage
	Curing Agent (Hardeners)	Curing Shrinkage Toughness Hardness Viscosity	Voids Incomplete Curing Non-uniform encapsulation Warpage Délamination
Connection	Metal Wire	Hardness Corrosion resistance	Corrosion Disconnection
	Micro-bump		Voids, Crack Defects Corrosion Disconnection Warpage Intermetallic compound formation

such as delamination, fracturing, or disconnection once they occur. A list of common failures is given in Table 7.2. After locating the point of failure, material and mechanical characterization methods will be applied to determine its root cause. Some in-process test methods will be performed again to assist in this effort. Once the cause of failure is understood, the material, structure, and chip fabrication process are adjusted to improve the reliability of the IC package.

7.2.2 Packaging Verification and Detection Methods

The IC package integrity assessment with in-processing testing and failure analysis can be performed in either non-destructive or destructive detection methods. The effectiveness of different detection methods varies widely depending on chip complexity and material composition, so multiple techniques must be leveraged in this verification process.

7.2.2.1   Non-destructive Detection

Several different non-destructive detection modalities are used in the industry. Some of the most common are explained below.

1. **X-Ray Photoelectron Spectroscopy (XPS):** In XPS, electrons are fired from an X-ray source onto the sample under test [39]. Semi-quantitative information about the elements inside the chip packaging can be extracted by measuring the kinetic energy of the ejected electrons.
2. **Scanning Electron Microscope Energy-Dispersive Spectroscopy (SEM-EDS):** When high-energy electrons are fired toward a sample, the collision results in X-rays leaving from the contact point. SEM-EDS characterizes this relationship to provide both qualitative and quantitative information about the materials that constitute a sample [31]. Since an SEM electron beam's diameter is very small, SEM-EDS can identify sample elements with a very high spatial resolution [40]. Its penetration depth is also deeper; this allows SEM-EDS to detect elements below the surface of the chip packaging.
3. **X-Ray Fluorescence (XRF):** XRF is another inspection method capable of material analysis. It can be performed on gasses, liquids, and solids, and it requires minimal sample preparation. Furthermore, its detection capabilities outperform SEM-EDS. It can determine material presence even if only 0.1% of the sample is composed of a given element [29].
4. **Scan Acoustic Microscopy (SAM):** SAM imaging can be applied to measure the composition of a chip package [28]. By analyzing reflected and transmitted ultrasound waves, SAM can non-destructively detect cracks and delamination inside an IC package. Cracks and defects which contain air or a vacuum will show as a dark sport on an SAM image. Because the amplitude and polarity of the ultrasound changes with inhomogeneities and discontinuities within a material, SAM can also characterize material uniformity. Unfortunately, the detection process can take up to several minutes to analyze one sample. As such, it is mainly useful when applied to random samples from a manufacturing run. Additionally, SAM resolution is not sufficient for detecting sub-micrometer micro-bumps, which are the connection between the die and interposer in advanced IC packaging [34].
5. **X-Ray Tomography:** X-ray tomography has been widely used for inspecting details and defects in packages by leveraging the difference in X-ray absorption between various packaging materials [33]. X-rays can penetrate samples without decapsulation, allowing internal structures to be extracted by 2D X-ray imaging [14]. 3D views of ICs are also capable through X-ray laminography, which combines 2D images of different layers. In all cases, X-ray imaging is an ideal tool to extract the subsurface structure of IC packaging. However, X-ray imaging quality depends on the algorithm used for imaging extrication, imaging time, and X-ray source quality [9]. Hence, it is only useful for real-time verification in structure characterization rather than full assurance.

In summary, non-destructive techniques can be quite effective for detecting material defects such as scratches, broken connections, and related faults. However, they cannot provide robust compositional analysis. As a result, destructive detection methods are required to address those concerns.

7.2.2.2 Destructive Detection

While non-destructive detection methods cannot provide rigorous assurance of complex packaging material, destructive methods such as molding compound characterization and internal examination can fit these needs. The techniques described below only partially address such concerns; more robust detection techniques are required for complete internal packaging characterization.

1. **Polymer Material Characterization:** Four basic approaches are widely used to characterize mechanical properties of encapsulant and underfill molding compound materials [20].

   a. **Differential Scanning Calorimeter (DSC):** DSC can measure the curing condition, which can help evaluate the hardness of packaging polymers.
   b. **Thermogravimetric Analysis (TGA):** TGA can analyze the glass transition temperature (Tg) of the polymer material by measuring sample weight loss under heat.
   c. **Thermomechanical Analysis (TMA):** TMA measures the expansion of the sample under a specific temperature, which is the coefficient of thermal expansion (CTE). CTE is analyzed while evaluating the stability of chip packaging.
   d. **Dynamic Mechanical Thermal Analysis (DMA):** DMA can measure the storage modulus of the encapsulating material, which is also used to verify the reliability of chip packaging under different stresses.

2. **Fourier-Transform Infrared Spectroscopy (FTIR) and Raman Spectroscopy:** These two are the most widely used package material chemical property characterization methods. By measuring the energy absorbed or scattered by different kinds of material, they can both qualitatively and quantitatively characterize the polymer material.

3. **Scanning Electron Microscopy (SEM):** SEM can be employed once an IC's packaging is removed through decapsulation [10]. For this reason, it is characterized as destructive analysis here. SEM is commonly used to detect defects at nanometer resolution. During SEM imaging, an electron beam is applied to a small area of the sample surface, generating several unique signals: secondary electrons, internal electrons, and photon emission. By analyzing these signals, different subsurface information can be extracted. Depending on the beam energy used during SEM imaging, materials will respond differently. Thus, material profiling is also possible through this method. SEM-EDS, while often used for non-destructive imaging, can also be applied during destructive testing. The

surface image created by SEM reveals the same information as in optical imaging but with much higher resolution and greater depth of field. The higher resolution of SEM can help detect much smaller defects inside chip packaging, and greater depth of field can be used to characterize the texture of the surface samples. With the help of stereo-photogrammetry technology, a 3D imaging of the sample surface can also be built [6]. 3D SEM has more depth information, which can help texture analysis of the sample surface.

4. **Interface Mechanical Reliability Testing:** Electronic interconnections between different IC layers are critical to IC package reliability. Wire-pull testing, ball shear testing, and die shear testing methods are used to directly measure the reliability of such interconnections. The details of such tests are covered in Sect. 7.4.

7.3 Vulnerability Assessment of IC Packaging

While traditional inspection techniques outlined in the previous section are effective at detecting most reliability issues related to packaging techniques, they are far less successful at detecting malicious packaging changes. The paragraphs below detail such threat models designed to avoid common inspection techniques while violating package security and reliability.

7.3.1 Threat Model

As shown in Fig. 7.4, there are four contributing factors to IC package reliability: design, process, environment, and material [29]. This discussion assumes design is controlled by the foundries and the environment can be controlled by the end users.

During production, designers will not deliberately violate package design rules to degrade or alter the packaging quality. Also, the user will follow the design rules and run the device under a typical environment. In other words, neither

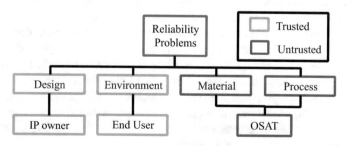

Fig. 7.4 Threat model of advanced IC packaging

party will deliberately decrease IC packaging reliability. The only two contributing factors controlled by the offshore fabricators are fabrication process and material, which can threaten IC packaging. However, any fabricator can change the materials or processes of the packaging for commercial or security purposes. Furthermore, complex ICs with advanced packaging are even more difficult to verify with existing methods due to the dense component placement and small feature sizes. As a result, these issues will become more pronounced over time. For these reasons, it is essential to understand the impact of material properties as they relate to IC package assurance.

7.3.2   Packaging Material Reliability Issues

Chip packaging security threats mainly arise from the difficulty inherent in assuring completed electronic devices. From the threat model described above, IC packaging properties will be changed when an untrusted fabricator alters the IC packaging process and material. The most common materials analyzed when determining packaging reliability are listed in Table 7.2 [6], along with the reliability problems they create.

In the following discussion, polymer package materials will be used as an example to explain the limitations of existing detection methods and the origin of packaging assurance problems. Critical encapsulant materials consist of epoxy resin matrices, fillers, and hardeners which control Tg, CTE, and mechanical properties of the IC packaging. Altering package composition through any of these parameters will likely cause chip failure. The mechanism of these reliability problems has been well studied, and it will not be covered here [6]. However, existing detection methods may exhibit limited effectiveness in determining packaging property changes caused by material or process differences. As a result, the outsourced fabricator can purposely cause reliability problems in these areas.

Underfill materials used in advanced IC packaging exhibit similar failure modes to encapsulants and hence enter the discussion as well. It is often injected between the die and interposer to reduce the CTE mismatch and increase thermal transportation (TC) of the packaging [27]. If any changes occur in the material composition or component, mechanical properties such as Tg, CTE, and TC could be changed, and the reliability problems that have listed above may result.

The ideal approach to preventing failures from encapsulant and underfill manipulation is to apply real environment loading on the chip samples and perform failure analysis afterward. However, these tests are time-consuming, and such a procedure will decrease the lifetime of the chips [24]. Hence, this approach is not viable for assurance purposes. Instead, in-process testing on the end-user side is implemented. Unfortunately, due to the disadvantages and shortcomings of non-decapsulated in-process detection methods, some reliability problems caused by material composition may remain undetected.

Since material defects can be implanted on a chip-by-chip basis, complete assurance cannot be achieved through random sampling alone. However, none of the methods shown in Sect. 7.2 can non-destructively evaluate the properties of all manufactured chips due to time constraints. Since, in theory, all referenced mechanical properties can be measured via non-contact means, another option for providing assurance is to measure the materials of each chip package and compare it with the golden samples. If the material composition matches within a predetermined tolerance, the chip packaging can be verified.

Fortunately, non-destructive methods such as XPS, XRF, and EDS can detect material composition. However, XPS and EDS are surface element analysis methods, offering detection only up to electron penetration depths. Additionally, XRF is not very sensitive to polymer elements such as carbon, hydrogen, and oxygen. Finally, these detection methods can only detect the atomic element itself and are quite time-consuming. Hence, if, for instance, the underfill inside an advanced chip package has been changed, it is difficult to know without decapsulating the chip package. Reliability issues stem from these limitations and must be addressed for ICs used in critical systems.

7.4 Non-destructive Bond Wire Integrity Assessment: A Case Study

The connectivity of packaging with die is the most research reliability topic in the packaging industry. Wire bonding is a commonly used approach to connect the die to the lead frame [35]. However, ball grid array (BGA) technologies are gaining popularity because they lower cost, reduce package size, and improve electrical and thermal performance [13]. An IC package is subjected to different electrical, thermal, and mechanical loads during its life cycle. Such adverse conditions may cause deformations or cracks in the bond wires and BGA. There have been extensive studies to achieve desired reliability and durability in bond wire and ball grid arrays [42]. Several destructive assessment methods have been proposed to detect defects in the bond wire and BGA. However, physical inspection techniques, such as X-ray tomography, can evaluate IC package integrity. In this section, we will discuss a non-destructive approach to bond wire integrity assessment.

Bond Pull and Ball Shear Test

The two most prevalent tests used to capture tolerance of chips to harsh environments are bond pull and ball shear [13, 32] (see Fig. 7.5). Bond pull testing is performed to evaluate the strength and quality of a wire bond. As the name suggests, the bond pull method requires the operator to apply an upward force under the wire under test using a small hook to pull it away from the die [19]. The bond shear test assesses the bonding between the base ball bond and the bonding pad. A chisel-shaped ram is positioned at the base of the ball and pushed against the base ball,

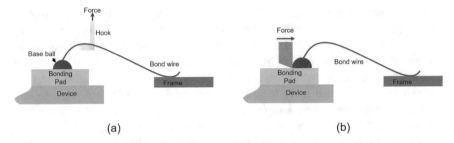

Fig. 7.5 Bond wire integrity analysis with (**a**) bond pull test and (**b**) ball shear test

breaking it off from the bonding pad [13]. Parameters from these tests are collected to assess bonding quality.

The testing tools used for the aforementioned tests require physical access to the bond wires. For this purpose, the chip must be decapsulated either locally or entirely. Decapsulation of packaging materials can be performed using chemicals such as acids. Acid can etch off entire packaging material, which sometimes can damage the interconnects. Local decapsulation is performed with laser- or plasma-based etching methods. Mechanical polishing is used to partially remove the packaging material for achieving better performance with etching. After decapsulation, destructive or non-destructive tests can be performed on bond wires and BGAs. None of the chips subjected to either of these test methods will be suitable for field use afterward [41, 43].

In a destructive test, force is applied on the bond wire until one of the following failures occurs: wire break, bond pad peeling, wire neck break, or ball shift. Ball shear testing is applicable for locating faults such as weld interface separation, partial metallization lift off, bond pad lift, or cratering [21, 38]. In the non-destructive test, however, the force will not exceed 90% of the designed strength of the bond wire [36, 37].

Decapsulation produces several disadvantages, regardless of whether the testing is destructive or not:

- The process is costly, destructive, and time-consuming and requires additional expertise. Moreover, polishing and chemical or plasma etching require dedicated equipment for decapsulation.
- Only one physical test can be applied on single bond wire or solder ball and each wire can be used for only one physical test.
- Fatigue analysis requires very long testing times and a controlled experiment environment.
- The effect of mechanical load (e.g., pull force and shear force) is exerted only on one wire at a time.

In recent years, there has been a significant effort to develop virtual methods—particularly finite element modeling (FEM)—to assess wire bond wire integrity [26, 37]. Although finite element analysis (FEA) offers a much faster testing procedure and unique testing environment, the solution requires accurate initial and boundary conditions. Besides, the wire bonding process and the epoxy molding process play a crucial role in the reliability of ICs. Such information can only be acquired from the geometry and structure of real test samples. As gaining physical access to bond wires is a costly and time-consuming process, the FEM is always a simplified version of the real model. Such limitations can be overcome with 3D X-ray tomography and image processing algorithms. Micro-CT X-ray can be used to develop a model of bond wire and ball grid arrays. Later, these models can be used in FEA to assess the stress distribution and displacement in bond wires. Such models include process variation, cracks, or voids, which play a crucial role in predicting the failure mode of the bond wire and the chip's reliability.

7.4.1 Methodology

An example of non-destructive bond wire integrity analysis is demonstrated in this section. The TL145406N microchip by Texas Instruments and the AD7512DIJN by Analog Devices were both evaluated. Figure 7.6 shows the chips used for analysis. However, the complete procedures for non-destructive bond pull and ball shear tests described here are independent of chip type.

7.4.2 X-Ray Tomography

Accurate information from the shape and geometry of test samples is required to create a finite element (FE) model of bond wires. X-ray computed tomography is one common method for gathering structural information of bond wires. In this example, the GE Phoenix micro-CT system was used to acquire the X-ray images. The GE Phoenix is a dual tube system capable of scanning high- and low-density material. The system is implemented with 180 kV nano-focus tube and a 240 kV

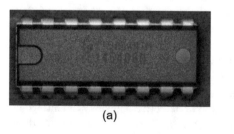

(a)   (b)

Fig. 7.6 Optical microscope image of (**a**) Texas Instruments TL145406N (chip 1) and (**b**) Analog Devices AD7512DIJN (chip 2)

Table 7.3 Tomography
parameters for microchips

Parameters	TL145406N	AD7512DIJN
Magnification	35.94	37.98
Voxel size (μm)	5.56	5.26
Focus-object-distance (mm)	22.39	21.2
Focus-detector-distance (mm)	805.22	805.22
Number of projections	1200	1200
Source voltage (kV)	200	200
Source current (μA)	25	25

Fig. 7.7 Non-destructive 3D
image of the test chips using
X-ray tomography (**a**) Texas
Instruments TL145406N
(chip 1) and (**b**) Analog
Devices AD7512DIJN
(chip 2)

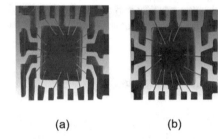

(a)   (b)

micro-focus tube, located inside the system. The tubes can move in an automated
fashion based on the required imaging resolution and sample type. The system
contains a dynamic DXR digital detector with 2024 × 2024 pixels. The tomography
parameters are defined based on empirical analysis to minimize the noise in the
image and achieve the highest transmission. The tomography parameters used for
this case study are provided in Table 7.3.

After the X-ray data was collected, Datos software from Phoenix was used
for 3D image reconstruction. The geometry and the internal structure of all bond
wires can be captured using non-destructive X-ray tomography. A major concern
in tomography is removing imaging artifacts. The reconstruction process and
equipment imperfections are the two most frequent sources of artifacts. Common
sources of error in these categories include beam hardening, scattering, and ring
artifacts. These are explained in greater detail below.

Common Artifacts of X-Ray Tomography

1. Beam hardening occurs due to the attenuation of low-energy photons when an
   X-ray beam passes through dense material. Attenuation of low-energy photons
   increases the mean beam energy and causes streaks or dark bands to appear in
   the image [11, 12]. To avoid such artifacts, pre-filters of metallic material like
   aluminum, copper, and brass can be used.
2. Due to scattering, X-ray photons can end up on different areas of detector where
   less photons are anticipated under standard conditions.
3. Ring artifacts are caused by a miscalibrated or defective detector element, which
   results in rings centered on the rotation origin. This can often be fixed by
   recalibrating the detector image [12].

As seen in Fig. 7.7, the geometry and the internal structure of all bond wires can be captured using non-destructive X-ray tomography. Material properties are available from the chip datasheet or original component manufacturer (OCM). The material properties and bond wire/ball grid geometry will be sufficient to create a finite element model and analyze chip integrity.

7.4.2.1 Image Segmentation

Next, the acquired 3D image from X-ray tomography is converted into a real finite element model. Once the FEM is ready for analysis, the model is processed with one of a number of available numerical solvers such as Autodesk SimStudio, Abaqus, COMSOL, Nastran, etc. This is referred to as the reverse engineering step of the process. The bond wires of the ICs presented in Fig. 7.6 are then used for FEA. The technique for the non-destructive bond pull test with FEA is discussed in this chapter. However, other types of similar tests such as thermomechanical tests, fatigue, multiphysics analysis, vibration, impact test, etc. can be modeled in the same way using different solver packages.

The acquired 3D images can be stored in .tiff, .jpg, or .pcr format. Since these formats are not suitable as a CAD model, image processing is used in this example to develop the bond wires' CAD model. VGStudio max 22 from Virtual Graphics group is used for post-processing the X-ray images. VG is a powerful image processing package and is widely used for image processing, filtering, segmentation, etc., on various types of samples, such as semiconductors and electronics, fossils, and more. The segmentation process is completed in three steps:

1. First, the region growing (flooding) algorithm is used to segment each bond wire separately in a semi-automated fashion based on the reconstructed 3D X-ray image histogram. Although this step can be done for all bond wires at once, the accuracy will be lower due to the noise (coming from beam hardening or center shift) in the images, which is common for most of CT images.
2. The bond wire is segmented according to the local gray-value gradients. This step eliminates sharp edges in the surface, which are mostly due to the noise in the image and are difficult to mesh during the finite element analysis.
3. Once the whole bond wire is segmented, the model is exported as a stereolithography (.stl) file, a meshed representation of a 3D standard format for CAD models which can be easily imported to FE packages or into a 3D printer. The number of vertices and elements in the mesh influences the accuracy and size of .stl file. This file describes only the triangulated surface geometry.
4. A stereolithography file does not contain any solids which are necessary for FE analysis. So, the .stl file is later converted into solid. Different software like Autodesk Inventor, Autodesk SimStudio tools, Ansys, SolidWorks, etc. can perform this task. Before the conversion into solid, mesh simplification can be applied on the .stl file.

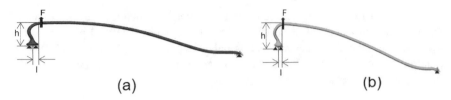

Fig. 7.8 Segmented structure with initial and boundary conditions and applied load direction for the FE analysis for (**a**) bond wire of TL145406N and (**b**) bond wire of AD7512DIJN

Mesh simplification or decimation is a process of reducing edge, vertex, or triangles while preserving the boundary and shape [15]. The accuracy of boundary and shape of simplified mesh depend on the algorithm and percentage of simplification. Mesh simplification can reduce the size of the CAD file and reduce the computational time for FE analysis. The solid model is finally converted into .stp or .step file format for FE analysis.

Once the .stp file is ready, it can be imported into Autodesk Simulation Mechanical, Autodesk Nastran In-CAD, or Ansys for further analysis or modification. The models of wire bonds are imported into .stp files into Autodesk Mechanical for further analysis. As shown in Fig. 7.8, geometrical properties of bond wires—such as loop part (l), loop height (h), and wire diameter—can also be measured virtually from FE model.

Limitations

Imaging resolution and detectability are the only limiting factors for the accuracy of the reverse-engineered model. Though industrial X-ray CT systems provide a wide range of imaging resolution from tens of nanometers up to millimeters, the optimum resolution is mostly dependent on the sample size. For example, the GE X-ray system can acquire high-resolution images down to 1 micron. Although this can detect multiple cracks causing failure in bond wires, any smaller cracks are likely to pass undetected. Ultra CT systems, which are not yet as popular as the micro-CT systems, can provide down to 50 nm resolution; however, the sample size needs to be minuscule and in the range of few hundreds of microns since the X-ray energy and power are only below 10 KV. Therefore, an ultra CT system is not currently viable for non-destructive bond wire analysis.

7.4.3   Finite Element Model Preparation

There are three different types of bond wire failures [8]:

1. Bond lift off from pad metallization on ball bond or stitch side.
2. Wire break at bond heel crack or neck break on ball bond or stitch side.
3. Wire break that occurs on any place other than the previous two cases.

As is evident from the description of the first type of bond wire failures, bond lift is about the reliability of the wire attached to the die and pad and not about the wire itself. Details about ball bond or the wire attachment to the die are currently very difficult to detect, especially in smaller technology nodes since the die thickness is below 1 μm and is too thin for the current X-ray technology to scan. This will be the only limitation of non-destructive assessment, though this is technically a limitation of the imaging and not of the proposed concept, which is the main focus of the paper.

Two bond wires from each chip are shown in Fig. 7.8. The CAD models of bond wires are imported into the Autodesk Simulation. For the material used to form the bond wires, copper for TL145406N and aluminum for AD7512DIJN were selected. The next important step for FE modeling is to mesh the model so that critical regions are covered with an optimized number of elements. Due to the irregular shape and geometry of bond wires, tetrahedral elements were selected for meshing. These are compatible with third-party finite element analysis packages. The four-node tetrahedral element also provides a uniform mesh for 3D analysis since such elements can easily sit next to each other and cover the whole area. There are no midside nodes included for this element, and the second-order integration is selected as default. Once wires are meshed, boundary conditions and loading can be applied. Figure 7.8 shows the location where the loading force of 1 cN is applied [19, 35]. Loading is applied at the peak of the wires since literature indicates this is where bond wires are most commonly tested for physical tests. Fixed boundary conditions on both ball bond and stitch-end sides are also added to the model. This means the model at this stage will only include the two types of failure mentioned earlier. Once the model is prepared, any FE solver including Autodesk Nastran, Autodesk Mechanical, or Ansys can be used.

7.4.4 Results

Physical bond pull and shear testing result in the failure of both bond wire and solder balls, which means the chip is no longer functional. There are standards for the loading force in both testing, depending on the chip type and size. Results in this section present the critical regions where the failure has the highest likelihood of occurrence. Knowing where failure is likely can allow optimization of a design or wiring process to minimize the critical regions.

The results for one wire from chips 1 and 2 are discussed here. Since both aluminum and copper are ductile materials, von Mises stress distribution and displacement magnitude are shown here. Figure 7.9 shows the von Mises stress distribution in the bond wires for chip 1 and chip 2, respectively. Stress estimation is calculated based on a stress-to-nodes method. This method provides greater accuracy by using an extrapolation scheme, as compared to stress calculations that are obtained by using stress inside elements. From Fig. 7.9, it is evident that two regions pointed as 1 and 2 show maximum stress value. This implies that the regions

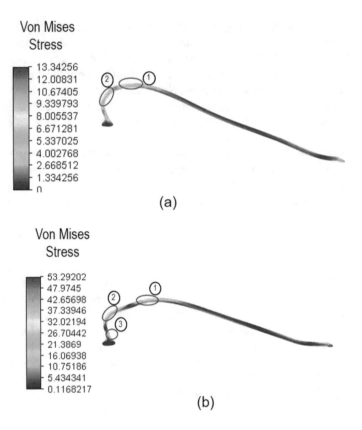

Fig. 7.9 von Mises stress distribution (N/mm^2) of bond wires for (**a**) TL145406N and (**b**) AD7512DIJN

marked as 1 and 2 in the figures are suspicious for early failure. From the bond wire of chip 1, a third region near ball bond end can be identified with a high stress value which could be due to the non-standard shape of the wire. In many cases, the bond wire is not generated in a standard form, which can cause such stress concentrations and may end to early failure. For the first chip, the maximum stress value is about 53 N/mm^2, and for the second chip, it is about 13 N/mm^2.

Figure 7.10 presents the displacement magnitude for both chips. Displacement magnitudes greater than zero show the total distance the node has moved. The maximum displacement is about 70 microns for the first chip and 3 microns for the second chip. Although stress distribution is more important for failure analysis, the displacement magnitude distribution confirms that failure is not a direct function of maximum displacement in bond wire problems.

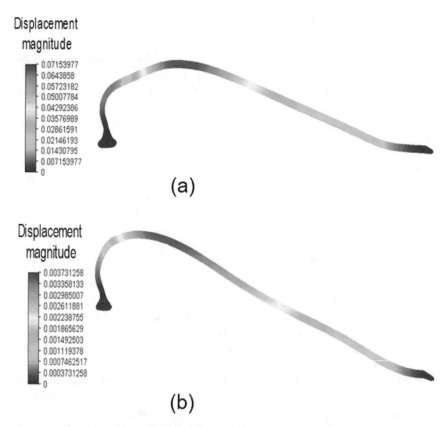

Displacement
magnitude

(a)

Displacement
magnitude

(b)

Fig. 7.10 Displacement magnitude distribution (mm) for (**a**) TL145406N and (**b**) AD7512DIJN

7.5   Requirements for IC Packaging Assurance

Previous sections have outlined the history of IC packaging and complexities therein, as well as the assurance challenges that have arisen over time and various solutions to combat malicious packaging and material alterations. However, these tests are still carried out on only a handful of manufactured samples and hence cannot be used to guarantee assurance for every fabricated chip. Here, the concept of "100% assurance" will be introduced. This is composed of three parts:

1. Defect detection must be performed on the whole chip package. No sample preparation steps can be employed due to time constraints and because each sample must be fully operational after verification.
2. The detection must be performed in real-time. Toward this end, scan times must be on the order of seconds, or multiple samples must be assured simultaneously.
3. To remain viable for fully automated workflows, this process should incorporate none or minimum human involvement.

7.5.1 Potential Solutions

As is the case in other assurance domains, fingerprinting negates the need for complete reverse engineering to verify material integrity. When determining what information should constitute a fingerprint, only fast and non-destructive methods such as physically unclonable function (PUF) tags [7] can apply due to the prior discussed constraints. These features have been widely used in the manufacturing market, as representative cases below describe:

- Diamond Unclonable Security Tag (DUST) is one such commercially available PUF [3]. Each diamond crystal on the samples has different crystallographic orientations. The combination of different orientations from different crystals generates a unique tag for that sample. Through detecting this device tag, a device can be traced during later stages of the supply chain.
- Molecule-level random variations, such as those present in polymers [25] and DNA [23], can also be used as an anti-counterfeit tag.

However, for package material assurance purposes, PUFs may not be very useful because the information of tag is not from the chip material itself. It can only verify the authentication of the devices—not the material inside the IC packaging

Instead, a standard tag generated by the standard IC packaging material is needed. With the help of the existing detection methods, this tag information can be extracted.

Fourier-transform infrared spectroscopy (FTIR) is one viable option to fill these needs. In this instance, wavenumbers are extracted, which represent different chemical function groups of polymer materials. These numbers are ideal tags since they relate directly to the IC packaging material. While this method requires some sample preparation, reflective mode FTIR can reduce the degree of preparation needed. It can also be quite time-consuming. However, the detection can be designed to extract several specific wavenumbers from the IC packaging material rather than all wavenumbers to speed up data retrieval. Finally, the nature of advanced IC packaging makes it quite difficult to analyze internal contents non-destructively [34].

Terahertz (THz) inspection is another attractive assurance method which does not require sample preparation. THz time-domain spectroscopy (TDS) has been used for anti-counterfeit fingerprinting purposes, as seen in other examples [4]. Another significant advantage of THz imaging is its transparency to silicon, meaning it can easily detect the polymer material inside a chip package [5], which solves the third reservation from FTIR tags. However, THz TDS can be quite sensitive to several noise sources. As such, it is critical to select a noise-invariant tag. This selection requires significant experimentation to prove that the tag is not affected by other parameters such as aging or moisture absorption. To further increase the speed and accuracy of the detection, the tag extraction process can be automated, reducing the manual work required for each sample.

References

1. Amkor Technology: https://amkor.com/
2. CCD Sensor Problems in Consumer Imaging Products: https://www.imaging-resource.com/badccds.html
3. DUST IDENTITY: https://dustidentity.com/
4. Ahi, K., Asadizanjani, N., Shahbazmohamadi, S., Tehranipoor, M., Anwar, M.: Terahertz characterization of electronic components and comparison of terahertz imaging with x-ray imaging techniques. Terahertz Physics, Devices, and Systems IX: Advanced Applications in Industry and Defense **9483**, 94830K (2015). https://doi.org/10.1117/12.2183128
5. Ahi, K., Shahbazmohamadi, S., Asadizanjani, N.: Quality control and authentication of packaged integrated circuits using enhanced-spatial-resolution terahertz time-domain spectroscopy and imaging. Opt. Lasers Eng. **104**, 274–284 (2018). https://doi.org/10.1016/j.optlaseng.2017.07.007
6. Ardebili, H., Zhang, J., Pecht, M.: Encapsulation technologies for electronic applications
7. Arppe, R., Sørensen, T.J.: Physical unclonable functions generated through chemical methods for anti-counterfeiting. Nat. Rev. Chem. **1** (2017). https://doi.org/10.1038/s41570-017-0031
8. Asadi, N., Forte, D., Tehranipoor, M.: Non-destructive bond pull and ball shear failure analysis based on real structural properties. In: 42nd International Symposium for Testing and Failure Analysis (ISTFA) (2016)
9. Asadizanjani, N., Tehranipoor, M., Forte, D.: PCB reverse engineering using nondestructive X-ray tomography and advanced image processing. IEEE Trans. Compon. Packag. Manuf. Technol. **7**(2), 292–299 (2017). https://doi.org/10.1109/TCPMT.2016.2642824
10. Bajura, M., Boverman, G., Tan, J., Wagenbreth, G., Rogers, C.M., Feser, M., Rudati, J., Tkachuk, A., Aylward, S., Reynolds, P.: Imaging integrated circuits with X-ray microscopy. In: Proceedings of 36th GOMACTech Conference (March), pp. 3–7 (2011)
11. Barrett, J.F., Keat, N.: Artifacts in ct: recognition and avoidance. Radiographics **24**(6), 1679–1691 (2004)
12. Boas, F.E., Fleischmann, D.: Ct artifacts: causes and reduction techniques. Imaging Med. **4**(2), 229–240 (2012)
13. Brunnbauer, M., Meyer, T., Ofner, G., Mueller, K., Hagen, R.: Embedded wafer level ball grid array (EWLB). In: 2008 33rd IEEE/CPMT International Electronics Manufacturing Technology Conference (IEMT), pp. 1–6. IEEE (2008)
14. Chen, Y., Lin, N., Lai, P.: Three-dimensional X-Ray laminography as a tool for detection and characterization of package on package(PoP) defects. In: ICRMS 2014 – Proceedings of 2014 10th International Conference on Reliability, Maintainability and Safety: More Reliable Products, More Secure Life, vol. 25(2), pp. 275–278 (2014). https://doi.org/10.1109/ICRMS.2014.7107186
15. Cignoni, P., Montani, C., Scopigno, R.: A comparison of mesh simplification algorithms. Comput. Graph. **22**(1), 37–54 (1998)
16. Bohr, M.: The new era of scaling in an SoC world. In: 2009 IEEE International Solid-State Circuits Conference-Digest of Technical Papers, pp. 23–28. IEEE (2009)
17. Fan, X.: Wafer level packaging (WLP): fan-in, fan-out and three-dimensional integration. In: 2010 11th International Thermal, Mechanical & Multi-Physics Simulation, and Experiments in Microelectronics and Microsystems (EuroSimE), pp. 1–7. IEEE (2010)
18. Guin, U., Huang, K., Dimase, D., Carulli, J.M., Tehranipoor, M., Makris, Y.: Counterfeit integrated circuits: a rising threat in the global semiconductor supply chain. Proc. IEEE **102**(8), 1207–1228 (2014). https://doi.org/10.1109/JPROC.2014.2332291
19. Harman, G., Cannon, C.: The microelectronic wire bond pull test-how to use it, how to abuse it. IEEE Trans. Compon. Hybrids Manuf. Technol. **1**(3), 203–210 (1978)
20. He, Y., Moreira, B.E., Overson, A., Nakamura, S.H., Bider, C., Briscoe, J.F.: Thermal characterization of an epoxy-based underfill material for flip chip packaging. Thermochim. Acta **357–358**, 1–8 (2000). https://doi.org/10.1016/S0040-6031(00)00357-9

21. Huang, X., Lee, S.W., Yan, C.C., Hui, S.: Characterization and analysis on the solder ball shear testing conditions. In: 2001 Proceedings. 51st Electronic Components and Technology Conference (Cat. No. 01CH37220), pp. 1065–1071. IEEE (2001)
22. Andrews, J.A.: Flip Chip Package and Method of Making (1994)
23. Kang, K., Park, J., Hyoung, S., Oh, J.Y.: (12) United States Patent 2(12) (2014)
24. Labie, R., Limaye, P., Lee, K.W., Berry, C.J., Beyne, E., De Wolf, I.: Reliability testing of Cu-Sn intermetallic micro-bump interconnections for 3D-device stacking. In: Electronics System Integration Technology Conference, ESTC 2010 – Proceedings (2010). https://doi.org/10.1109/ESTC.2010.5642925
25. Laure, C., Karamessini, D., Milenkovic, O., Charles, L., Lutz, J.F.: Coding in 2D: using intentional dispersity to enhance the information capacity of sequence-coded polymer barcodes. Angewandte Chemie – International Edition 55(36), 10722–10725 (2016). https://doi.org/10.1002/anie.201605279
26. Liu, Y., Irving, S., Luk, T.: Thermosonic wire bonding process simulation and bond pad over active stress analysis. In: 2004 Proceedings. 54th Electronic Components and Technology Conference (IEEE Cat. No. 04CH37546), vol. 1, pp. 383–391. IEEE (2004)
27. Lu, D., Wong, C.P.: Materials for Advanced Packaging, 2nd edn. Springer International Publishing (2016). https://doi.org/10.1007/978-3-319-45098-8
28. Ma, L., Bao, S., Lv, D., Du, Z., Li, S.: Application of C-mode scanning acoustic microscopy in packaging. In: Proceedings of the Electronic Packaging Technology Conference, EPTC 00, 3–8 (2007). https://doi.org/10.1109/ICEPT.2007.4441498
29. Moore, T.: Characterization of integrated circuit packaging materials. Elsevier, Amsterdam (2013)
30. Mounttechnology, S., Smith, I.W.D., Dennis, R., Richard, C.: United States Patent (19) (19) (1989)
31. Newbury, D.E., Ritchie, N.W.: Is scanning electron microscopy/energy dispersive X-ray spectrometry (SEM/EDS) quantitative? Scanning 35(3), 141–168 (2013). https://doi.org/10.1002/sca.21041
32. Price, S.F., Munakata, H., Razon, E., Perlberg, G., Fokin, I.: Diagnostic wire bond pull tester (1997). US Patent 5,591,920
33. Shahbazmohamadi, S., Forte, D., Tehranipoor, M.: Advanced physical inspection methods for counterfeit IC detection. In: Conference Proceedings from the International Symposium for Testing and Failure Analysis, 55–64 (2014)
34. Tu, K.N.: Reliability challenges in 3D IC packaging technology. Microelectron. Reliab. 51(3), 517–523 (2011). https://doi.org/10.1016/j.microrel.2010.09.031
35. Van Driel, W., Janssen, J., Van Silfhout, R., Van Gils, M., Zhang, G., Ernst, L.: On wire failures in micro-electronic packages. In: Proceedings of the 5th International Conference on Thermal and Mechanical Simulation and Experiments in Microelectronics and Microsystems, 2004. EuroSimE 2004, pp. 53–57. IEEE (2004)
36. Van Gils, M., van der Sluis, O., Zhang, G., Janssen, J., Voncken, R.: Analysis of cu/low-k bond pad delamination by using a novel failure index. In: EuroSimE 2005. Proceedings of the 6th International Conference on Thermal, Mechanial and Multi-Physics Simulation and Experiments in Micro-Electronics and Micro-Systems, 2005, pp. 190–196. IEEE (2005)
37. Viswanath, A.G., Fang, W., Zhang, X., Ganesh, V., Lim, L.: Numerical analysis by 3d finite element wire bond simulation on cu/low-k structures. In: 2005 7th Electronic Packaging Technology Conference, vol. 1, p. 6. IEEE (2005)
38. Wang, C., Sun, R.: The quality test of wire bonding. Modern Appl. Sci. 3(12), 50–56 (2009)
39. Watts, J.F., Wolstenholme, J.: An Introduction to Surface Analysis by XPS and AES. Wiley, Chichester (2003). https://doi.org/10.1002/0470867930
40. Wilson, R., Asadizanjani, N., Forte, D., Woodard, D.L.: Histogram-Based Auto Segmentation: A Novel Approach to Segmenting Integrated Circuit Structures from SEM Images, pp. 4–9
41. Yeh, C.L., Lai, Y.S., Kao, C.L.: Transient simulation of wire pull test on cu/low-k wafers. IEEE Trans. Adv. Packag. 29(3), 631–638 (2006)

42. Zahn, B.A.: Finite element based solder joint fatigue life predictions for a same die size-stacked-chip scale-ball grid array package. In: 27th Annual IEEE/SEMI International Electronics Manufacturing Technology Symposium, pp. 274–284. IEEE (2002)
43. Zhang, X., Teysseyre, J., Goh, K.Y., Wong, W.: Copper wirebond pull test and reliability characterization with finite element simulation. In: 2011 IEEE 13th Electronics Packaging Technology Conference, pp. 20–24. IEEE (2011)

Appendix A
Failure Analysis Tools

A.1 Leica DVM6

A high-resolution digital microscope, such as Leica DVM6 (see Fig. A.1), may have zoom ratio of 16:1. The digital microscope has a fully apochromatic corrected optics and a 10M pixel camera with fast live imaging mode. The motorized stage has 3 degrees of freedom in x, y, and z direction which enables automatic stitching and 3D surface imaging. Since the Mark and Find software is available on this system, the user can define multiple stage locations and revisit them automatically during experiment setup or as part of time-lapse experiment. Such capabilities are very beneficial for a variety of hardware security projects including integrated circuits' counterfeit detection, PCB reverse engineering, wafers' failure analysis, etc.

A.2 TESCAN SEM Microscope

There are different types of scanning electron microscopes (SEM). Here we will discuss FERA-3 and LYRA-3 SEM microscope manufactured by TESCAN.

A.2.1 FERA-3

The FERA-3 (see Fig. A.2) is a fully PC controlled SEM with Schottky field emission cathode in combination with Xe plasma focused ion beam (i-FIB) column. This dual beam is also equipped with gas injection system (GIS) which enables depositing conductive and non-conductive material. The ion optics is a fully integrated Xe plasma ion source with 2 pA–2 μA probe current range and 3–30 Kv beam energy range. This probe will enable the milling rates to be typically 50–

N. Asadizanjani et al., *Physical Assurance*, https://doi.org/10.1007/978-3-030-62609-9

Fig. A.1 Leica DVM6
digital microscope

Fig. A.2 FERA-3 SEM
microscope

60 times faster than a conventional Ga liquid metal ion source (LIMIS) FIB.
Imaging is based on a high-performance DSP-based image acquisition system in
this microscope, which is capable of acquiring images up to 16k × 16k pixels with
a 16-bit dynamic range (pixel depth), and it can continuously adjust the scanning
rates from 20 ns per pixel to 10 ms per pixel to achieve a better signal-to-noise ratio.

A.2.2   LYRA-3

Similar to FERA-3, LYRA-3 (see Fig. A.3) is also fully PC controlled SEM with
Schottky field emission cathode in combination with a gallium i-FIB column.
Wide-field optics design in this system is available with an ultimate resolution of
1.2 nm. The ion column is a fully integrated Canion FIB column with Ga LIMIS,
with 1 pa–40 nA probe current range, and 0.5–30 kv beam energy, which provides
better control of material depositing or milling of a fine size item. The system
also has a fully integrated XYZ piezo nano-manipulator with integrated closed-loop
nanoposition sensors and controllers for TEM sample preparation. The gas injection
system is equipped with automatic nozzle positioning and automatic temperature
control. Available precursors for deposition are carbon, platinum, tungsten, and
silicon oxide plus fluorine and water for etching.

Fig. A.3 LYRA-3 SEM microscope

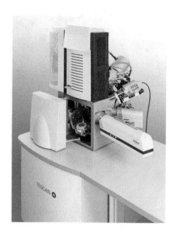

A.3 ORION NanoFab

It is a multi-ion column system including helium and neon that provides a complete sub-10 nm nano-fabrication and sub-nanometer imaging solution for research in a wide range of applications. The ORION NanoFab (see Fig. A.4) is a microscope with the abovementioned capabilities. The imaging resolution with helium ion is 0.5 nm, which is not achievable by regular Ga ion FIBs or SEMs. ORION NanoFab features the NanoPatterning and Visualization Engine (NVPE)—an integrated hardware and software control system. NVPE incorporates a dedicated 16-bit scan generator for each NanoFab column and dual signal acquisition hardware supporting real-time advanced patterning and visualization. It allows users to completely control the beam by a GUI and create a range of fully editable shapes including rectangles, trapezoids, polygons, lines, polylines, ellipses, and spots. The next step allows these shapes to be vector filled while maintaining full control over dose variation and patterning parameters.

A.4 PHEMOS-1000

The PHEMOS-1000 (see Fig. A.5) is a high-resolution optical emission microscope in near-infrared (NIR) spectrum that localizes failures in semiconductor devices by detecting the light emissions caused by semiconductor device defects. Moreover, its laser scan system allows acquiring high-resolution pattern images. Different types of detectors are available for various analysis techniques, such as photon emission analysis (PEM), electro-optical probing (EOP), electro-optical frequency mapping (EOFM), laser fault injection (LFI), and optical beam induced resistance change (OBIRCH) analysis. In addition to failure analysis applications, this microscope can

Fig. A.4 ORION NanoFab
used for sub-nanometer
imaging

Fig. A.5 PHEMOS-1000
with spectrum analyzer and
laser source

be deployed for reverse engineering of integrated circuits (ICs) from the backside
of the package.

A.5 SkyScan 2211

The X-ray nano-CT system (see Fig. A.6) covers the largest range of objects sizes
and spatial resolution available in one instrument. The system has capacity for
scanning and 3D non-destructive reconstruction of internal micro-structures on
object sizes greater than 8 inches (200 mm, such as PCBs) as well as sub-micron
resolution for small samples (such as microchips). It is an "open-type" (pumped)
X-ray source, 20–190Kv with standard Be window, with sub-micron spot size
and two X-ray detectors: flat panel for large objects and 11 Mpixel cooled CCD
for scanning with the highest resolution down to 100 nm. It automatically adjusts

Fig. A.6 SkyScan 2211
nano-CT system used for
X-ray imaging

Fig. A.7 Summit 12000B
probe station used for
microprobing

for variable acquisition geometry, and phase contrast enhancement allows the best
possible quality in relatively short scanning times.

A.6 Summit 12000B Probe Station

The Summit 12000B semi-automated probe station (see Fig. A.7) is a flexible test
instrument for electronic devices including 200 and 150 mm wafers. Applications
of the instrument includes RF/Microwave, device characterization, reliability, modeling, and much more. Our model is equipped with an ATT thermal system as
well as an EzLaze 3 Nd:YAD Infrared Laser Cutting System. The measurement
environment is encased within an electromagnetic shielding and light-inhibiting
enclosure. Security applications for this equipment include integrated circuit aging
analysis, device level and wafer characterization, PCB inspection and modification,
and related high precision spatial measurements.

A.7 Nanoprober

Imina Technologies' Nanoprobing SEM Solutions (see Fig. A.8) are turnkey for
electrical characterization of microelectronic devices and in situ semiconductor

Fig. A.8 miBot robot used
for nanoprobing

failure analysis. Up to 8 miBot™ nanoprobers can be delivered in various configurations to adapt to customer applications' requirements and equipment. The miBot
is a famously easy-to-use and versatile piezo actuated microrobot that allows you
to position the probes over millimeter-scale samples with a resolution down to the
nanometer. The 4 degrees of freedom of these nanoprobers enables the operator to
easily adjust the orientation of probes in situ during experiment.

Specifically designed for low current measurements, electrical characterization
of nanostructures can be carried out with third-party source meter units (SMU) and
signal analyzers through the shielded cabling, featuring an excellent signal-to-noise
ratio.

Equipped with the EBIC/EBAC, in situ pre-amplifiers and scan generators are
compatible with the nanoprobing solutions to perform quantitative EBIC and low-
noise EBAC/RCI analyses as well.

Index

Printed in the United States
by Baker & Taylor Publisher Services